Introduction to Earthquake Engineering

Introduction to Earthquake Engineering

Elijah Walker

STATES
ACADEMIC PRESS
www.statesacademicpress.com

States Academic Press,
109 South 5th Street,
Brooklyn, NY 11249, USA

Visit us on the World Wide Web at:
www.statesacademicpress.com

ISBN: 978-1-63989-303-4 (Hardback)

Cataloging-in-Publication Data

Introduction to earthquake engineering / Elijah Walker.
p. cm.
Includes bibliographical references and index.
ISBN 978-1-63989-303-4
1. Earthquake engineering. 2. Engineering geology. 3. Engineering. I. Walker, Elijah.
QE522 .I58 2022
624.176 2--dc23

Table of Contents

Preface

This book has been written, keeping in view that students want more practical information. Thus, my aim has been to make it as comprehensive as possible for the readers. I would like to extend my thanks to my family and co-workers for their knowledge, support and encouragement all along.

The branch of engineering that is concerned with the designing and analysis of structures to avoid damages caused by earthquakes is known as earthquake engineering. The goal of the discipline is to make structures like buildings and bridges that are resistant to earthquakes. It aims to make structures that do not stumble in case of minor shaking or collapse during a major earthquake. It focuses on the protection of natural and man-made environment and societies from earthquakes by stopping risks to acceptable levels. Earthquake engineering has two primary objectives; to foresee the consequences of earthquakes and design and maintain structures that can sustain earthquakes. The field is considered a subset of mechanical engineering, chemical engineering, applied physics, geotechnical engineering and structural engineering. This book unfolds the innovative aspects related to this field which will be crucial for the holistic understanding of the subject matter. It also elucidates new techniques and their applications in a multidisciplinary approach. This book is meant for students who are looking for an elaborate reference text on earthquake engineering.

A brief description of the chapters is provided below for further understanding:

Chapter - What is Earthquake Engineering?

Earthquake is the shaking of the Earth's surface due to the passage of seismic waves through Earth's rocks. Earthquake engineering is the branch of engineering that deals with the designing, analysis and building of earthquake-resistant structures. This is an introductory chapter which will briefly introduce about earthquake engineering.

Chapter - Earthquake Measurement

The magnitude of earthquake can be measured through the use of seismometers, seismograph, Richter magnitude scale, accelerograph, modified Mercalli intensity scale, environmental seismic intensity scale, etc. The topics elaborated in this chapter will help in gaining a better perspective of these instruments used in measuring earthquake.

Chapter - Seismology: A Comprehensive Study

Seismology refers to the study of earthquakes and seismic waves. Seismic gap, seismic migration, seismic noise, seismo-electromagnetics, paleoseismology, etc. are studied under its domain. All these aspects related to seismology have been carefully analyzed in this chapter.

Chapter – Seismic Hazard Assessment

Seismic hazard assessment refers to the effort for prediction of potential ground shaking intensity from future earthquakes. It uses a common approach of probabilistic seismic hazard analysis in which the expected ground motions are evaluated at a given site. This chapter discusses the subject of seismic hazard assessment in detail.

Chapter – Seismic Protection

Seismic protection is referred to as the sophisticated design approach involved in the construction to minimize the damages to structure during earthquake. It includes seismic base isolation, seismic loading, seismic retrofit, seismic metamaterial, viscoelastic and fluid viscous damper. This chapter closely examines seismic protection to provide an extensive understanding of the subject.

Elijah Walker

1

What is Earthquake Engineering?

Earthquake is the shaking of the Earth's surface due to the passage of seismic waves through Earth's rocks. Earthquake engineering is the branch of engineering that deals with the designing, analysis and building of earthquake-resistant structures. This is an introductory chapter which will briefly introduce about earthquake engineering.

EARTHQUAKE

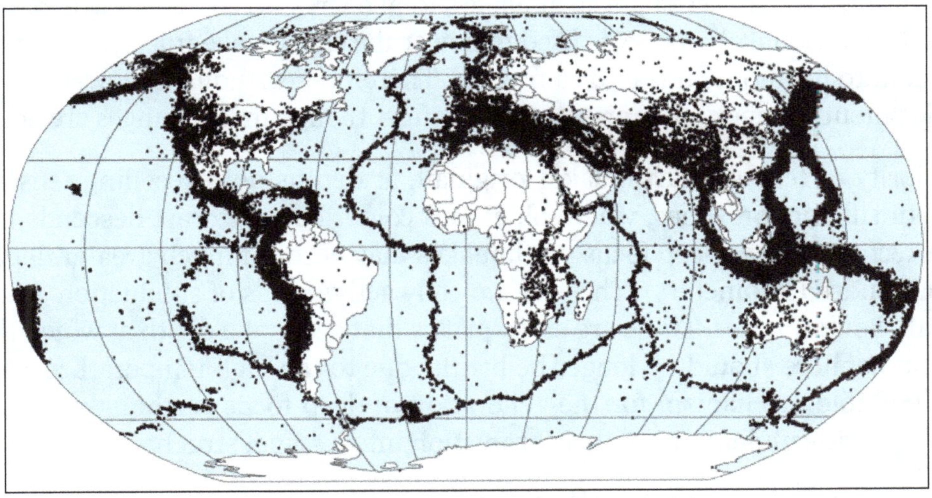

Global earthquake epicenters.

An earthquake is the result of a sudden release of stored energy in the Earth's crust that creates seismic waves. At the Earth's surface, earthquakes may manifest themselves by a shaking or displacement of the ground. Sometimes, they cause tsunamis, which may lead to loss of life and destruction of property. An earthquake is caused by tectonic plates getting stuck and putting a strain on the ground. The strain becomes so great that rocks give way by breaking and sliding along fault planes. Earthquakes are measured with a seismometer, commonly known as a seismograph.

Earthquakes may occur naturally or as a result of human activities. Smaller earthquakes can also be caused by volcanic activity, landslides, mine blasts, and nuclear tests. In its most generic sense, the word *earthquake* is used to describe any seismic event—whether a natural phenomenon or an event caused by humans—that generates seismic waves.

An earthquake's point of initial ground rupture is called its focus or hypocenter. The term epicenter means the point at ground level directly above this.

Naturally Occurring Earthquakes

Most naturally occurring earthquakes are related to the tectonic nature of the Earth. Such earthquakes are called *tectonic earthquakes*. The Earth's lithosphere is a patchwork of plates in slow but constant motion caused by the release to space of the heat in the Earth's mantle and core. The heat causes the rock in the Earth to become flow on geological timescales, so that the plates move, slowly but surely. Plate boundaries lock as the plates move past each other, creating frictional stress. When the frictional stress exceeds a critical value, called *local strength*, a sudden failure occurs. The boundary of tectonic plates along which failure occurs is called the *fault plane*. When the failure at the fault plane results in a violent displacement of the Earth's crust, the elastic strain energy is released and seismic waves are radiated, thus causing an earthquake. This process of strain, stress, and failure is referred to as the Elastic-rebound theory. It is estimated that only 10 percent or less of an earthquake's total energy is radiated as seismic energy. Most of the earthquake's energy is used to power the earthquake fracture growth and is converted into heat, or is released to friction. Therefore, earthquakes lower the Earth's available potential energy and raise its temperature, though these changes are negligible.

The majority of tectonic earthquakes originate at depths not exceeding tens of kilometers. In subduction zones, where older and colder oceanic crust descends beneath another tectonic plate, Deep focus earthquakes may occur at much greater depths (up to seven hundred kilometers). These seismically active areas of subduction are known as Wadati-Benioff zones. These are earthquakes that occur at a depth at which the subducted lithosphere should no longer be brittle, due to the high temperature and pressure. A possible mechanism for the generation of deep focus earthquakes is faulting caused by olivine undergoing a phase transition into a spinel structure.

Earthquakes may also occur in volcanic regions and are caused there both by tectonic faults and by the movement of magma in volcanoes. Such earthquakes can be an early warning of volcanic eruptions.

A recently proposed theory suggests that some earthquakes may occur in a sort of earthquake storm, where one earthquake will trigger a series of earthquakes each triggered by the previous shifts on the fault lines, similar to aftershocks, but occurring years later, and with some of the later earthquakes as damaging as the early ones. Such a pattern was observed in the sequence of about a dozen earthquakes that struck the North Anatolian Fault in Turkey in the twentieth century, the half dozen large earthquakes in New Madrid in 1811-1812, and has been inferred for older anomalous clusters of large earthquakes in the Middle East and in the Mojave Desert.

Size and Frequency of Occurrence

Small earthquakes occur nearly constantly around the world in places like California and Alaska in the United States, as well as in Chile, Indonesia, Iran, the Azores in Portugal, New Zealand, Greece, and Japan. Large earthquakes occur less frequently, the relationship being exponential; for example, roughly ten times as many earthquakes larger than magnitude 4 occur in a particular time period than earthquakes larger than magnitude 5. In the (low seismicity) United Kingdom, for example, it has been calculated that the average recurrences are:

- An earthquake of 3.7 or larger every year.
- An earthquake of 4.7 or larger every 10 years.
- An earthquake of 5.6 or larger every 100 years.

The number of seismic stations has increased from about 350 in 1931 to many thousands today. As a result, many more earthquakes are reported than in the past because of the vast improvement in instrumentation (not because the number of earthquakes has increased). The USGS estimates that, since 1900, there have been an average of 18 major earthquakes (magnitude 7.0-7.9) and one great earthquake (magnitude 8.0 or greater) per year, and that this average has been relatively stable. In fact, in recent years, the number of major earthquakes per year has actually decreased, although this is likely a statistical fluctuation. More detailed statistics on the size and frequency of earthquakes is available from the USGS.

Most of the world's earthquakes (90 percent, and 81 percent of the largest) take place in the 40,000-km-long, horseshoe-shaped zone called the circum-Pacific seismic belt, also known as the Pacific Ring of Fire, which for the most part bounds the Pacific Plate. Massive earthquakes tend to occur along other plate boundaries, too, such as along the Himalayan Mountains.

Effects/Impacts of Earthquakes

There are many effects of earthquakes including, but not limited to the following:

Shaking and Ground Rupture

Shaking and ground rupture are the main effects created by earthquakes, principally resulting in more or less severe damage to buildings or other rigid structures. The severity of the local effects depends on the complex combination of the earthquake magnitude, the distance from epicenter, and the local geological and geomorphological conditions, which may amplify or reduce wave propagation. The ground-shaking is measured by ground acceleration.

Specific local geological, geomorphological, and geostructural features can induce high levels of shaking on the ground surface even from low-intensity earthquakes. This effect is called site or local amplification. It is principally due to the transfer of the seismic motion from hard deep soils to soft superficial soils and to effects of seismic energy focalization owing to typical geometrical setting of the deposits.

Landslides and Avalanches

Earthquakes can cause landslides and avalanches, which may cause damage in hilly and mountainous areas.

Fires

Following an earthquake, fires can be generated by break of the electrical power or gas lines.

Soil Liquefaction

Soil liquefaction occurs when, because of the shaking, water-saturated granular material temporarily loses their strength and transforms from a solid to a liquid. Soil liquefaction may cause rigid structures, as buildings or bridges, to tilt or sink into the liquefied deposits.

Human Impacts

Earthquakes may result in disease, lack of basic necessities, loss of life, higher insurance premiums, general property damage, road and bridge damage, and collapse of buildings or destabilization of the base of buildings which may lead to collapse in future earthquakes.

Preparation for Earthquakes

- Emergency preparedness.
- Household seismic safety.
- HurriQuake nail (for resisting hurricanes and earthquakes).
- Seismic retrofit.
- Seismic hazard.
- Mitigation of seismic motion.
- Earthquake prediction.

Specific Fault Articles

- Alpine Fault.
- Calaveras Fault.

- Cascadia subduction zone.
- Geology of the Death Valley area.
- Great Glen Fault.
- Great Sumatran fault.
- Hayward Fault Zone.
- Highland Boundary Fault.
- Hope Fault.
- Liquiñe-Ofqui Fault.
- North Anatolian Fault Zone.
- New Madrid Fault Zone.
- San Andreas Fault.

Causes of Earthquakes

Earth's major earthquakes occur mainly in belts coinciding with the margins of tectonic plates. This has long been apparent from early catalogs of felt earthquakes and is even more readily discernible in modern seismicity maps, which show instrumentally determined epicentres. The most important earthquake belt is the Circum-Pacific Belt, which affects many populated coastal regions around the Pacific Ocean—for example, those of New Zealand, New Guinea, Japan, the Aleutian Islands, Alaska, and the western coasts of North and South America. It is estimated that 80 percent of the energy presently released in earthquakes comes from those whose epicentres are in this belt. The seismic activity is by no means uniform throughout the belt, and there are a number of branches at various points. Because at many places the Circum-Pacific Belt is associated with volcanic activity, it has been popularly dubbed the "Pacific Ring of Fire."

A second belt, known as the Alpide Belt, passes through the Mediterranean region eastward through Asia and joins the Circum-Pacific Belt in the East Indies. The energy released in earthquakes from this belt is about 15 percent of the world total. There also are striking connected belts of seismic activity, mainly along oceanic ridges—including those in the Arctic Ocean, the Atlantic Ocean, and the western Indian Ocean—and along the rift valleys of East Africa. This global seismicity distribution is best understood in terms of its plate tectonic setting.

Natural Forces

Earthquakes are caused by the sudden release of energy within some limited region of the rocks of the Earth. The energy can be released by elastic strain, gravity, chemical reactions, or even the motion of massive bodies. Of all these the release of elastic strain is the most important cause, because this form of energy is the only kind that can be stored in sufficient quantity in the Earth to produce major disturbances. Earthquakes associated with this type of energy release are called tectonic earthquakes.

Tectonics

Tectonic earthquakes are explained by the so-called elastic rebound theory, formulated by the American geologist Harry Fielding Reid after the San Andreas Fault ruptured in 1906, generating the great San Francisco earthquake. According to the theory, a tectonic earthquake occurs when strains in rock masses have accumulated to a point where the resulting stresses exceed the strength of the rocks, and sudden fracturing results. The fractures propagate rapidly through the rock, usually tending in the same direction and sometimes extending many kilometres along a local zone of weakness. In 1906, for instance, the San Andreas Fault slipped along a plane 430 km (270 miles) long. Along this line the ground was displaced horizontally as much as 6 metres (20 feet).

As a fault rupture progresses along or up the fault, rock masses are flung in opposite directions and thus spring back to a position where there is less strain. At any one point this movement may take place not at once but rather in irregular steps; these sudden slowings and restartings give rise to the vibrations that propagate as seismic waves. Such irregular properties of fault rupture are now included in the modeling of earthquake sources, both physically and mathematically. Roughnesses along the fault are referred to as asperities, and places where the rupture slows or stops are said to be fault barriers. Fault rupture starts at the earthquake focus, a spot that in many cases is close to 5–15 km under the surface. The rupture propagates in one or both directions over the fault plane until stopped or slowed at a barrier. Sometimes, instead of being stopped at the barrier, the fault rupture recommences on the far side; at other times the stresses in the rocks break the barrier, and the rupture continues.

Earthquakes have different properties depending on the type of fault slip that causes them. The usual fault model has a "strike" (that is, the direction from north taken by a horizontal line in the fault plane) and a "dip" (the angle from the horizontal shown by the steepest slope in the fault). The lower wall of an inclined fault is called the footwall. Lying over the footwall is the hanging wall. When rock masses slip past each other parallel to the strike, the movement is known as strike-slip faulting. Movement parallel to the dip is called dip-slip faulting. Strike-slip faults are right lateral or left lateral, depending on whether the block on the opposite side of the fault from an observer has moved to the right or left. In dip-slip faults, if the hanging-wall block moves downward relative to the footwall block, it is called "normal" faulting; the opposite motion, with the hanging wall moving upward relative to the footwall, produces reverse or thrust faulting.

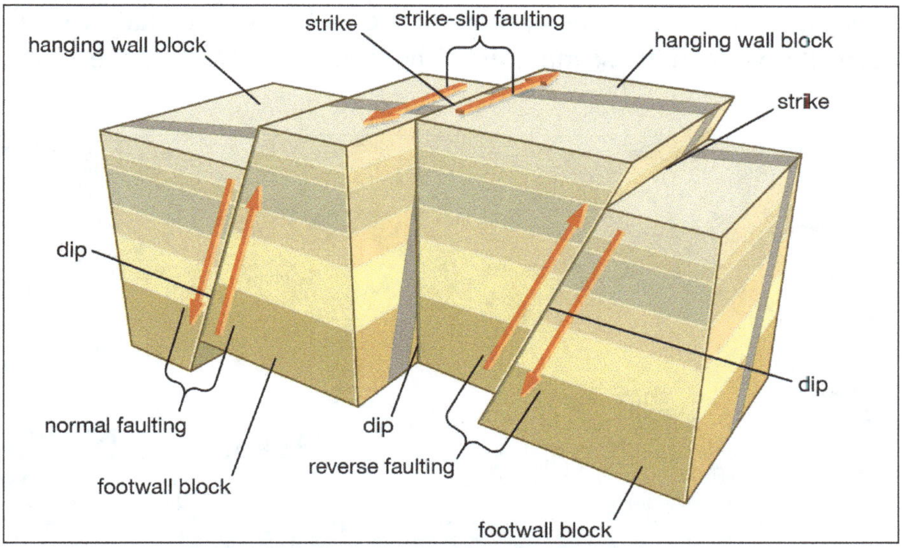

Types of faulting in tectonic earthquakesIn normal and reverse faulting, rock masses slip vertically past each other. In strike-slip faulting, the rocks slip past each other horizontally.

All known faults are assumed to have been the seat of one or more earthquakes in the past, though tectonic movements along faults are often slow, and most geologically ancient faults are now aseismic (that is, they no longer cause earthquakes). The actual faulting associated with an earthquake may be complex, and it is often not clear whether in a particular earthquake the total energy issues from a single fault plane.

Observed geologic faults sometimes show relative displacements on the order of hundreds of kilometres over geologic time, whereas the sudden slip offsets that produce seismic waves may range from only several centimetres to tens of metres. In the 1976 Tangshan earthquake, for example, a surface strike-slip of about one metre was observed along the causative fault east of Beijing, and in the 1999 Taiwan earthquake the Chelung-pu fault slipped up to eight metres vertically.

Volcanism

A separate type of earthquake is associated with volcanic activity and is called a volcanic earthquake. Yet it is likely that even in such cases the disturbance is the result of a sudden slip of rock masses adjacent to the volcano and the consequent release of elastic strain energy. The stored energy, however, may in part be of hydrodynamic origin due to heat provided by magma moving in reservoirs beneath the volcano or to the release of gas under pressure.

There is a clear correspondence between the geographic distribution of volcanoes and major earthquakes, particularly in the Circum-Pacific Belt and along oceanic ridges. Volcanic vents, however, are generally several hundred kilometres from the epicentres of most major shallow earthquakes, and many earthquake sources occur nowhere near active volcanoes. Even in cases where an earthquake's focus occurs directly below structures marked by volcanic vents, there is probably no immediate causal connection between the two activities; most likely both are the result of the same tectonic processes.

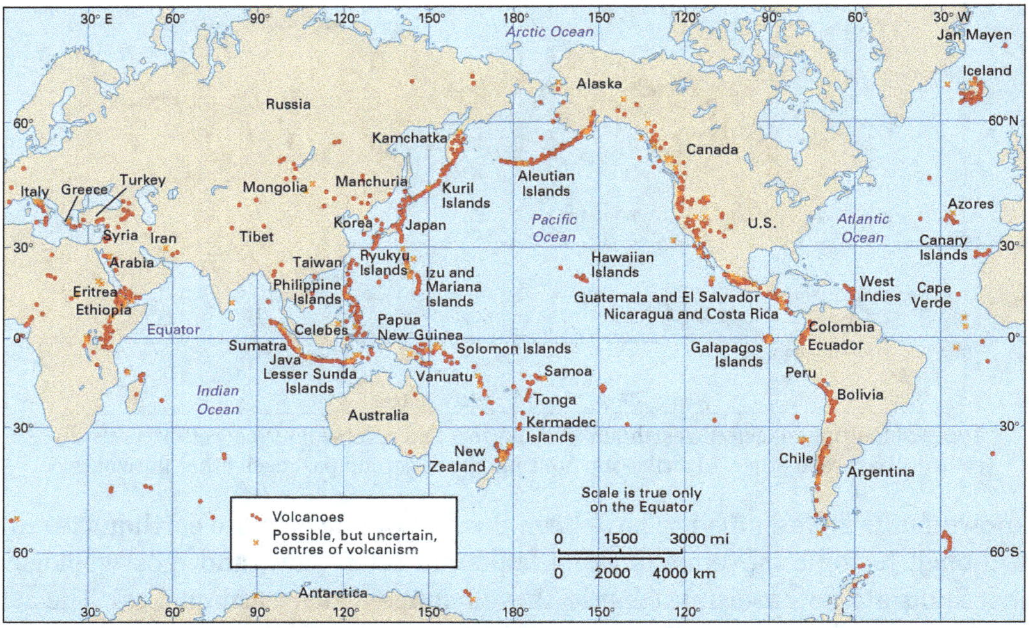

Volcanoes and thermal fields that have been active during the past 10,000 years.

Artificial Induction

Earthquakes are sometimes caused by human activities, including the injection of fluids into deep wells, the detonation of large underground nuclear explosions, the excavation of mines, and the filling of large reservoirs. In the case of deep mining, the removal of rock produces changes in the strain around the tunnels. Slip on adjacent, preexisting faults or outward shattering of rock into the new cavities may occur. In fluid injection, the slip is thought to be induced by premature release of elastic strain, as in the case of tectonic earthquakes, after fault surfaces are lubricated by the liquid. Large underground nuclear explosions have been known to produce slip on already strained faults in the vicinity of the test devices.

Reservoir Induction

Of the various earthquake-causing activities cited above, the filling of large reservoirs is among the most important. More than 20 significant cases have been documented in which local seismicity has increased following the impounding of water behind high dams. Often, causality cannot be substantiated, because no data exists to allow comparison of earthquake occurrence before and after the reservoir was filled. Reservoir-induction effects are most marked for reservoirs exceeding 100 metres (330 feet) in depth and 1 cubic km (0.24 cubic mile) in volume. Three sites where such connections have very probably occurred are the Hoover Dam in the United States, the Aswan High Dam in Egypt, and the Kariba Dam on the border between Zimbabwe and Zambia. The most generally accepted explanation for earthquake occurrence in such cases assumes that rocks near the reservoir are already strained from regional tectonic forces to a point where nearby faults are almost ready to slip. Water in the reservoir adds a pressure perturbation that triggers the fault rupture. The pressure effect is perhaps enhanced by the fact that the rocks along the fault have lower strength because of increased water-pore pressure. These factors notwithstanding, the filling of most large reservoirs has not produced earthquakes large enough to be a hazard.

The specific seismic source mechanisms associated with reservoir induction have been established in a few cases. For the main shock at the Koyna Dam and Reservoir in India, the evidence favours strike-slip faulting motion. At both the Kremasta Dam in Greece and the Kariba Dam in Zimbabwe-Zambia, the generating mechanism was dip-slip on normal faults. By contrast, thrust mechanisms have been determined for sources of earthquakes at the lake behind Nurek Dam in Tajikistan. More than 1,800 earthquakes occurred during the first nine years after water was impounded in this 317-metre-deep reservoir in 1972, a rate amounting to four times the average number of shocks in the region prior to filling.

Seismology and Nuclear Explosions

In 1958 representatives from several countries, including the United States and the Soviet Union, met to discuss the technical basis for a nuclear test-ban treaty. Among the matters considered was the feasibility of developing effective means with which to detect underground nuclear explosions and to distinguish them seismically from earthquakes. After that conference, much special research was directed to seismology, leading to major advances in seismic signal detection and analysis.

Recent seismological work on treaty verification has involved using high-resolution seismographs in a worldwide network, estimating the yield of explosions, studying wave attenuation in the Earth, determining wave amplitude and frequency spectra discriminants, and applying seismic arrays. The findings of such research have shown that underground nuclear explosions, compared with natural earthquakes, usually generate seismic waves through the body of the Earth that are of much larger amplitude than the surface waves. This telltale difference along with other types of seismic evidence suggest that an international monitoring network of 270 seismographic stations could detect and locate all seismic events over the globe of magnitude 4 and above (corresponding to an explosive yield of about 100 tons of TNT).

Intensity and Magnitude of Earthquakes

Intensity Scales

The violence of seismic shaking varies considerably over a single affected area. Because the entire range of observed effects is not capable of simple quantitative definition, the strength of the shaking is commonly estimated by reference to intensity scales that describe the effects in qualitative terms. Intensity scales date from the late 19th and early 20th centuries, before seismographs capable of accurate measurement of ground motion were developed. Since that time, the divisions in these scales have been associated with measurable accelerations of the local ground shaking. Intensity depends, however, in a complicated way not only on ground accelerations but also on the periods and other features of seismic waves, the distance of the measuring point from the source, and the local geologic structure. Furthermore, earthquake intensity, or strength, is distinct from earthquake magnitude, which is a measure of the amplitude, or size, of seismic waves as specified by a seismograph reading.

A number of different intensity scales have been set up during the past century and applied to both current and ancient destructive earthquakes. For many years the most widely used was a 10-point scale devised in 1878 by Michele Stefano de Rossi and Francois-Alphonse Forel. The scale now generally employed in North America is the Mercalli scale, as modified by Harry O. Wood and Frank Neumann in 1931, in which intensity is considered to be more suitably graded. A 12-point abridged form of the modified Mercalli scale is provided below. Modified Mercalli intensity VIII is roughly correlated with peak accelerations of about one-quarter that of gravity (g = 9.8 metres, or 32.2 feet, per second squared) and ground velocities of 20 cm (8 inches) per second. Alternative scales have been developed in both Japan and Europe for local conditions. The European (MSK) scale of 12 grades is similar to the abridged version of the Mercalli.

Modified Mercalli Scale of Earthquake Intensity

- Not felt. Marginal and long-period effects of large earthquakes.
- Felt by persons at rest, on upper floors, or otherwise favourably placed to sense tremors.
- Felt indoors. Hanging objects swing. Vibrations are similar to those caused by the passing of light trucks. Duration can be estimated.
- Vibrations are similar to those caused by the passing of heavy trucks (or a jolt similar to that caused by a heavy ball striking the walls). Standing automobiles rock. Windows, dishes, doors rattle. Glasses clink, crockery clashes. In the upper range of grade IV, wooden walls and frames creak.
- Felt outdoors; direction may be estimated. Sleepers awaken. Liquids are disturbed, some spilled. Small objects are displaced or upset. Doors swing, open, close. Pendulum clocks stop, start, change rate.
- Felt by all; many are frightened and run outdoors. Persons walk unsteadily. Pictures fall off walls. Furniture moves or overturns. Weak plaster and masonry cracks. Small bells ring (church, school). Trees, bushes shake.

- Difficult to stand. Noticed by drivers of automobiles. Hanging objects quivering. Furniture broken. Damage to weak masonry. Weak chimneys broken at roof line. Fall of plaster, loose bricks, stones, tiles, cornices. Waves on ponds; water turbid with mud. Small slides and caving along sand or gravel banks. Large bells ringing. Concrete irrigation ditches damaged.

- Steering of automobiles affected. Damage to masonry; partial collapse. Some damage to reinforced masonry; none to reinforced masonry designed to resist lateral forces. Fall of stucco and some masonry walls. Twisting, fall of chimneys, factory stacks, monuments, towers, elevated tanks. Frame houses moved on foundations if not bolted down; loose panel walls thrown out. Decayed pilings broken off. Branches broken from trees. Changes in flow or temperature of springs and wells. Cracks in wet ground and on steep slopes.

- General panic. Weak masonry destroyed; ordinary masonry heavily damaged, sometimes with complete collapse; reinforced masonry seriously damaged. Serious damage to reservoirs. Underground pipes broken. Conspicuous cracks in ground. In alluvial areas, sand and mud ejected; earthquake fountains, sand craters.

- Most masonry and frame structures destroyed with their foundations. Some well-built wooden structures and bridges destroyed. Serious damage to dams, dikes, embankments. Large landslides. Water thrown on banks of canals, rivers, lakes, and so on. Sand and mud shifted horizontally on beaches and flat land. Railway rails bent slightly.

- Rails bent greatly. Underground pipelines completely out of service.

- Damage nearly total. Large rock masses displaced. Lines of sight and level distorted. Objects thrown into air.

With the use of an intensity scale, it is possible to summarize such data for an earthquake by constructing isoseismal curves, which are lines that connect points of equal intensity. If there were complete symmetry about the vertical through the earthquake's focus, isoseismals would be circles with the epicentre (the point at the surface of the Earth immediately above where the earthquake originated) as the centre. However, because of the many unsymmetrical geologic factors influencing intensity, the curves are often far from circular. The most probable position of the epicentre is often assumed to be at a point inside the area of highest intensity. In some cases, instrumental data verify this calculation, but not infrequently the true epicentre lies outside the area of greatest intensity.

Earthquake Magnitude

Earthquake magnitude is a measure of the "size," or amplitude, of the seismic waves generated by an earthquake source and recorded by seismographs. (The types and nature of these waves are described in the section Seismic waves.) Because the size of earthquakes varies enormously, it is necessary for purposes of comparison to compress the range of wave amplitudes measured on seismograms by means of a mathematical device. In 1935 the American seismologist Charles F. Richter set up a magnitude scale of earthquakes as the logarithm to base 10 of the maximum seismic wave amplitude (in thousandths of a millimetre) recorded on a standard seismograph (the Wood-Anderson torsion pendulum seismograph) at a distance of 100 km (60 miles) from the earthquake epicentre. Reduction of amplitudes observed at various distances to the amplitudes expected at the standard distance of 100 km is made on the basis of empirical tables. Richter magnitudes M_L are computed on the assumption that the ratio of the maximum wave amplitudes at two given distances is the same for all earthquakes and is independent of azimuth.

Richter first applied his magnitude scale to shallow-focus earthquakes recorded within 600 km of the epicentre in the southern California region. Later, additional empirical tables were set up, whereby observations made at distant stations and on seismographs other than the standard type could be used. Empirical tables were extended to cover earthquakes of all significant focal depths and to enable independent magnitude estimates to be made from body- and surface-wave observations. A current form of the Richter scale is shown in the table.

Table: Richter scale of earthquake magnitude.

Magnitude level	Category	Effects	Earthquakes per year
Less than 1.0 to 2.9	Micro	Generally not felt by people, though recorded on local instruments	More than 100,000
3.0–3.9	Minor	Felt by many people; no damage	12,000–100,000
4.0–4.9	Light	Felt by all; minor breakage of objects	2,000–12,000
5.0–5.9	Moderate	Some damage to weak structures	200–2,000
6.0–6.9	Strong	Moderate damage in populated areas	20–200
7.0–7.9	Major	Serious damage over large areas; loss of life	3–20
8.0 and higher	Great	Severe destruction and loss of life over large areas	Fewer than 3

At the present time a number of different magnitude scales are used by scientists and engineers as a measure of the relative size of an earthquake. The *P*-wave magnitude (M_b), for one, is defined in terms of the amplitude of the *P* wave recorded on a standard seismograph. Similarly, the surface-wave magnitude (M_s) is defined in terms of the logarithm of the maximum amplitude of ground motion for surface waves with a wave period of 20 seconds.

As defined, an earthquake magnitude scale has no lower or upper limit. Sensitive seismographs can record earthquakes with magnitudes of negative value and have recorded magnitudes up to about 9.0. (The 1906 San Francisco earthquake, for example, had a Richter magnitude of 8.25.)

A scientific weakness is that there is no direct mechanical basis for magnitude as defined above. Rather, it is an empirical parameter analogous to stellar magnitude assessed by astronomers. In modern practice a more soundly based mechanical measure of earthquake size is used—namely, the seismic moment (M_0). Such a parameter is related to the angular leverage of the forces that produce the slip on the causative fault. It can be calculated both from recorded seismic waves and from field measurements of the size of the fault rupture. Consequently, seismic moment provides a more uniform scale of earthquake size based on classical mechanics. This measure allows a more scientific magnitude to be used called moment magnitude (M_w). It is proportional to the logarithm of the seismic moment; values do not differ greatly from M_s values for moderate earthquakes. Given the above definitions, the great Alaska earthquake of 1964, with a Richter magnitude (M_L) of 8.3, also had the values $M_s = 8.4$, $M_0 = 820 \times 10^{27}$ dyne centimetres, and $M_w = 9.2$.

Earthquake Energy

Energy in an earthquake passing a particular surface site can be calculated directly from the recordings of seismic ground motion, given, for example, as ground velocity. Such recordings indicate an energy rate of 10^5 watts per square metre (9,300 watts per square foot) near a moderate-size earthquake source. The total power output of a rupturing fault in a shallow earthquake is on the order of 10^{14} watts, compared with the 10^5 watts generated in rocket motors.

The surface-wave magnitude M_s has also been connected with the surface energy E_s of an earthquake by empirical formulas. These give $E_s = 6.3 \times 10^{11}$ and 1.4×10^{25} ergs for earthquakes of $M_s = 0$ and 8.9, respectively. A unit increase in M_s corresponds to approximately a 32-fold increase in energy. Negative magnitudes M_s correspond to the smallest instrumentally recorded earthquakes, a magnitude of 1.5 to the smallest felt earthquakes, and one of 3.0 to any shock felt at a distance of up to 20 km (12 miles). Earthquakes of magnitude 5.0 cause light damage near the epicentre; those of 6.0 are destructive over a restricted area; and those of 7.5 are at the lower limit of major earthquakes.

The total annual energy released in all earthquakes is about 10^{25} ergs, corresponding to a rate of work between 10 million and 100 million kilowatts. This is approximately one one-thousandth the annual amount of heat escaping from the Earth's interior. Ninety percent of the total seismic energy comes from earthquakes of magnitude 7.0 and higher—that is, those whose energy is on the order of 10^{23} ergs or more.

Frequency

There also are empirical relations for the frequencies of earthquakes of various magnitudes. Suppose N to be the average number of shocks per year for which the magnitude lies in a range about M_s. Then $\log_{10} N = a - bM_s$ fits the data well both globally and for particular regions; for example, for shallow earthquakes worldwide, $a = 6.7$ and $b = 0.9$

when $M_s > 6.0$. The frequency for larger earthquakes therefore increases by a factor of about 10 when the magnitude is diminished by one unit. The increase in frequency with reduction in M_s falls short, however, of matching the decrease in the energy E. Thus, larger earthquakes are overwhelmingly responsible for most of the total seismic energy release. The number of earthquakes per year with $M_b > 4.0$ reaches 50,000.

Occurrence of Earthquakes

Tectonic Associations

Global seismicity patterns had no strong theoretical explanation until the dynamic model called plate tectonics was developed during the late 1960s. This theory holds that the Earth's upper shell, or lithosphere, consists of nearly a dozen large, quasi-stable slabs called plates. The thickness of each of these plates is roughly 80 km (50 miles). The plates move horizontally relative to neighbouring plates at a rate of 1 to 10 cm (0.4 to 4 inches) per year over a shell of lesser strength called the asthenosphere. At the plate edges where there is contact between adjoining plates, boundary tectonic forces operate on the rocks, causing physical and chemical changes in them. New lithosphere is created at oceanic ridges by the upwelling and cooling of magma from the Earth's mantle. The horizontally moving plates are believed to be absorbed at the ocean trenches, where a subduction process carries the lithosphere downward into the Earth's interior. The total amount of lithospheric material destroyed at these subduction zones equals that generated at the ridges.

Seismological evidence (such as the location of major earthquake belts) is everywhere in agreement with this tectonic model. Earthquake sources are concentrated along the oceanic ridges, which correspond to divergent plate boundaries. At the subduction zones, which are associated with convergent plate boundaries, intermediate- and deep-focus earthquakes mark the location of the upper part of a dipping lithosphere slab. The focal mechanisms indicate that the stresses are aligned with the dip of the lithosphere underneath the adjacent continent or island arc.

Some earthquakes associated with oceanic ridges are confined to strike-slip faults, called transform faults, that offset the ridge crests. The majority of the earthquakes occurring along such horizontal shear faults are characterized by slip motions. Also in agreement with the plate tectonics theory is the high seismicity encountered along the edges of plates where they slide past each other. Plate boundaries of this kind, sometimes called fracture zones, include the San Andreas Fault in California and the North Anatolian fault system in Turkey. Such plate boundaries are the site of interplate earthquakes of shallow focus.

The low seismicity within plates is consistent with the plate tectonic description. Small to large earthquakes do occur in limited regions well within the boundaries of plates; however, such intraplate seismic events can be explained by tectonic mechanisms other than plate boundary motions and their associated phenomena.

Shallow, Intermediate and Deep Foci

Most parts of the world experience at least occasional shallow earthquakes—those that originate within 60 km (40 miles) of the Earth's outer surface. In fact, the great majority of earthquake foci are shallow. It should be noted, however, that the geographic distribution of smaller earthquakes is less completely determined than more severe quakes, partly because the availability of relevant data is dependent on the distribution of observatories.

Of the total energy released in earthquakes, 12 percent comes from intermediate earthquakes—that is, quakes with a focal depth ranging from about 60 to 300 km. About 3 percent of total energy comes from deeper earthquakes. The frequency of occurrence falls off rapidly with increasing focal depth in the intermediate range. Below intermediate depth the distribution is fairly uniform until the greatest focal depths, of about 700 km (430 miles), are approached.

The deeper-focus earthquakes commonly occur in patterns called Benioff zones that dip into the Earth, indicating the presence of a subducting slab. Dip angles of these slabs average about 45°, with some shallower and others nearly vertical. Benioff zones coincide with tectonically active island arcs such as Japan, Vanuatu, Tonga, and the Aleutians, and they are normally but not always associated with deep ocean trenches such as those along the South American Andes. Exceptions to this rule include Romania and the Hindu Kush mountain system. In most Benioff zones, intermediate- and deep-earthquake foci lie in a narrow layer, although recent precise hypocentral locations in Japan and elsewhere show two distinct parallel bands of foci 20 km apart.

Aftershocks, Foreshocks and Swarms

Usually, a major or even moderate earthquake of shallow focus is followed by many lesser-size earthquakes close to the original source region. This is to be expected if the fault rupture producing a major earthquake does not relieve all the accumulated strain energy at once. In fact, this dislocation is liable to cause an increase in the stress and strain at a number of places in the vicinity of the focal region, bringing crustal rocks at certain points close to the stress at which fracture occurs. In some cases an earthquake may be followed by 1,000 or more aftershocks a day.

Sometimes a large earthquake is followed by a similar one along the same fault source within an hour or perhaps a day. An extreme case of this is multiple earthquakes. In most instances, however, the first principal earthquake of a series is much more severe than the aftershocks. In general, the number of aftershocks per day decreases with time. The aftershock frequency is roughly inversely proportional to the time since the occurrence of the largest earthquake of the series.

Most major earthquakes occur without detectable warning, but some principal earthquakes are preceded by foreshocks. In another common pattern, large numbers of small earthquakes may occur in a region for months without a major earthquake. In the Matsushiro region of Japan, for instance, there occurred between August 1965 and August 1967 a series of hundreds of thousands of earthquakes, some sufficiently strong (up to Richter magnitude 5) to cause property damage but no casualties. The maximum frequency was 6,780 small earthquakes on April 17, 1966. Such series of earthquakes are called earthquake swarms. Earthquakes associated with volcanic activity often occur in swarms, though swarms also have been observed in many nonvolcanic regions.

Observation of Earthquakes

Earthquake Observatories

Worldwide during the late 1950s, there were only about 700 seismographic stations, which were equipped with seismographs of various types and frequency responses. Few instruments were calibrated; actual ground motions could not be measured, and timing errors of several seconds were common. The World-Wide Standardized Seismographic Network (WWSSN), the first modern worldwide standardized system, was established to help remedy this situation. Each station of the WWSSN had six seismographs—three short-period and three long-period seismographs. Timing and accuracy were maintained by crystal clocks, and a calibration pulse was placed daily on each record. By 1967 the WWSSN consisted of about 120 stations distributed over 60 countries. The resulting data provided the basis for significant advances in research on earthquake mechanisms, global tectonics, and the structure of the Earth's interior.

By the 1980s a further upgrading of permanent seismographic stations began with the installation of digital equipment by a number of organizations. Among the global networks of digital seismographic stations now in operation are the Seismic Research Observatories in boreholes 100 metres (330 feet) deep and modified high-gain, long-period surface observatories. The Global Digital Seismographic Network in particular has remarkable capability, recording all motions from Earth tides to microscopic ground motions at the level of local ground noise. At present there are about 128 sites. With this system the long-term seismological goal will have been accomplished to equip global observatories with seismographs that can record every small earthquake anywhere over a broad band of frequencies.

Locating Earthquake Epicentres

Many observatories make provisional estimates of the epicentres of important earthquakes. These estimates provide preliminary information locally about particular earthquakes and serve as first approximations for the calculations subsequently made by large coordinating centres.

If an earthquake's epicentre is less than 105° away from an observatory, the epicentre's position can often be estimated from the readings of three seismograms recording perpendicular components of the ground motion. For a shallow earthquake the epicentral distance is indicated by the interval between the arrival times of *P* and *S* waves; the azimuth and angle of wave emergence at the surface are indicated by a comparison of the sizes and directions of the first movements shown in the seismograms and by the relative sizes of later waves, particularly surface waves. It should be noted, however, that in certain regions the first wave movement at a station arrives from a direction differing from the azimuth toward the epicentre. This anomaly is usually explained by strong variations in geologic structures.

When data from more than one observatory are available, an earthquake's epicentre may be estimated from the times of travel of the *P* and *S* waves from source to recorder. In many seismically active regions, networks of seismographs with telemetry transmission and centralized timing and recording are common. Whether analog or digital recording is used, such integrated systems greatly simplify observatory work: multichannel signal displays make identification and timing of phase onsets easier and more reliable. Moreover, online microprocessors can be programmed to pick automatically, with some degree of confidence, the onset of a significant common phase, such as *P*, by correlation of waveforms from parallel network channels. With the aid of specially designed computer programs, seismologists can then locate distant earthquakes to within about 10 km (6 miles) and the epicentre of a local earthquake to within a few kilometres.

Catalogs of earthquakes felt by humans and of earthquake observations have appeared intermittently for many centuries. The earliest known list of instrumentally recorded earthquakes with computed times of origin and epicentres is for the period 1899–1903. In subsequent years, cataloging of earthquakes has become more uniform and complete. Especially valuable is the service provided by the International Seismological Centre (ISC) at Newbury, Eng. Each month it receives more than 1,000,000 readings from more than 2,000 stations worldwide and preliminary estimates of the locations of approximately 1,600 earthquakes from national and regional agencies and observatories. The ISC publishes a monthly bulletin—with about a two-year delay—that provides all available information on each of more than 5,000 earthquakes.

Various national and regional centres control networks of stations and act as intermediaries between individual stations and the international organizations. Examples of long-standing national centres include the Japan Meteorological Agency and United States National Earthquake Information Center in Colorado (a subdivision of the United States Geological Survey). These centres normally make estimates of the magnitudes, epicentres, origin times, and focal depths of local earthquakes. On the Internet, data on global seismicity is continually accessible through the Web site of the Incorporated Research Institutions for Seismology (IRIS).

An important research technique is to infer the character of faulting in an earthquake from the recorded seismograms. For example, observed distributions of the directions of the first onsets in waves arriving at the Earth's surface have been effectively used. Onsets are called "compressional" or "dilatational" according to whether the direction is away from or toward the focus, respectively. A polarity pattern becomes recognizable when the directions of the *P*-wave onsets are plotted on a map—there are broad areas in which the first onsets are predominantly compressions, separated from predominantly dilatational areas by nodal curves near which the *P*-wave amplitudes are abnormally small.

In 1926 the American geophysicist Perry E. Byerly used patterns of *P* onsets over the entire globe to infer the orientation of the fault plane in a large earthquake. The polarity method yields two *P*-nodal curves at the Earth's surface; one curve is in the plane containing the assumed fault, and the other is in the plane (called the auxiliary plane) that passes through the focus and is perpendicular to the forces of the plane. The recent availability of worldwide broad-based digital recording has enabled computer programs to be written that estimate the fault mechanism and seismic moment from the complete pattern of seismic wave arrivals. Given a well-determined pattern at a number of earthquake observatories, it is possible to locate two planes, one of which is the plane containing the fault.

Earthquake Prediction

Observation and Interpretation of Precursory Phenomena

The search for periodic cycles in earthquake occurrence is an old one. Generally, periodicities in time and space for major earthquakes have not been widely detected or accepted. One problem is that long-term earthquake catalogs are not homogeneous in their selection and reporting. The most extensive catalog of this kind comes from China and begins about 700 BCE. The catalog contains some information on about 1,000 destructive earthquakes. The sizes of these earthquakes have been assessed from the reports of damage, intensity, and shaking.

Another approach to the statistical occurrence of earthquakes involves the postulation of trigger forces that initiate the rupture. Such forces have been attributed to severe weather conditions, volcanic activity, and tidal forces, for example. Usually correlations are made between the physical phenomena assumed to provide the trigger and the repetition of earthquakes. Inquiry must always be made to discover whether a causative link is actually present, but in no cases to the present has a trigger mechanism, at least for moderate to large earthquakes, been unequivocally found that satisfies the various necessary criteria.

Statistical methods also have been tried with populations of regional earthquakes. It has been suggested, but never established generally, that the slope *b* of the regression line between the logarithm of the number of earthquakes and the magnitude for a region may change characteristically with time. Specifically, the claim is that the *b* value for the population of foreshocks of a major earthquake may be significantly smaller than the mean *b* value for the region averaged over a long interval of time.

The elastic rebound theory of earthquake sources allows rough prediction of the occurrence of large shallow earthquakes. Harry F. Reid gave, for example, a crude forecast of the next great earthquake near San Francisco. (The theory also predicted, of course, that the place would be along the San Andreas or an associated fault.) The geodetic data indicated that during an interval of 50 years relative displacements of 3.2 metres (10.5 feet) had occurred at distant points across the fault. The maximum elastic-rebound offset along the fault in the 1906 earthquake was 6.5 metres. Therefore, (6.5 ÷ 3.2) × 50, or about 100, years would again elapse before sufficient strain accumulated for the occurrence of an earthquake comparable to that of 1906. The premises are that the regional strain will grow uniformly and that various constraints have not been altered by the great 1906 rupture itself (such as by the onset of slow fault slip). Such strain rates are now being more adequately measured along a number of active faults such as the San Andreas, using networks of GPS sensors.

For many years prediction research has been influenced by the basic argument that strain accumulates in the rock masses in the vicinity of a fault and results in crustal deformation. Deformations have been measured in the horizontal direction along active faults (by trilateration and triangulation) and in the vertical direction by precise leveling and tiltmeters. Some investigators believe that changes in groundwater level occur prior to earthquakes; variations of this sort have been reported mainly from China. Because water levels in wells respond to a complex array of factors such as rainfall, such factors will have to be removed if changes in water level are to be studied in relation to earthquakes.

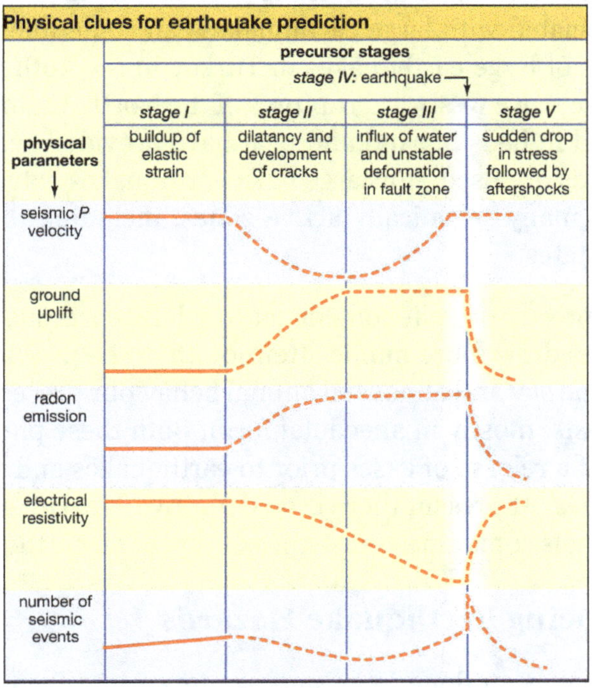

The theory of dilatancy (that is, an increase in volume) of rock prior to rupture once occupied a central position in discussions of premonitory phenomena of earthquakes, but it now receives less support. It is based on the observation that many solids exhibit dilatancy during deformation. For earthquake prediction the significance of dilatancy, if real, is in its effects on various measurable quantities of the Earth's crust, such as seismic velocities, electric resistivity, and ground and water levels. The consequences of dilatancy for earthquake prediction are summarized in the table. The best-studied consequence is the effect on seismic velocities. The influence of internal cracks and pores on the elastic properties of rocks can be clearly demonstrated in laboratory measurements of those properties as a function of hydrostatic pressure. In the case of saturated rocks, experiments predict—for shallow earthquakes—that dilatancy occurs as a portion of the crust is stressed to failure, causing a decrease in the velocities of seismic waves. Recovery of velocity is brought about by subsequent rise of the pore pressure of water, which also has the effect of weakening the rock and enhancing fault slip.

Strain buildup in the focal region may have measurable effects on other observable properties, including electrical conductivity and gas concentration. Because the electrical conductivity of rocks depends largely on interconnected water channels within the rocks, resistivity may increase before the cracks become saturated. As pore fluid is expelled from the closing cracks, the local water table would rise and concentrations of gases such as radioactive radon would increase. No unequivocal confirming measurements have yet been published.

Geologic methods of extending the seismicity record back from the present also are being explored. Field studies indicate that the sequence of surface ruptures along major active faults associated with large earthquakes can sometimes be constructed. An example is the series of large earthquakes in Turkey in the 20th century, which were caused mainly by successive westward ruptures of the North Anatolian Fault. Liquefaction effects preserved in beds of sand and peat have provided evidence—when radiometric dating methods are used—for large paleoearthquakes extending back for more than 1,000 years in many seismically active zones, including the Pacific Northwest coast of the United States.

Less well-grounded precursory phenomena, particularly earthquake lights and animal behaviour, sometimes draw more public attention than the precursors . Many reports of unusual lights in the sky and abnormal animal behaviour preceding earthquakes are known to seismologists, mostly in anecdotal form. Both these phenomena are usually explained in terms of a release of gases prior to earthquakes and electric and acoustic stimuli of various types. At present there is no definitive experimental evidence to support claims that animals sometimes sense the coming of an earthquake.

Methods of Reducing Earthquake Hazards

Considerable work has been done in seismology to explain the characteristics of the recorded ground motions in earthquakes. Such knowledge is needed to predict ground motions in future earthquakes so that earthquake-resistant structures can be designed. Although earthquakes cause death and destruction through such secondary effects as landslides, tsunamis, fires, and fault rupture, the greatest losses—both of lives and of property—result from the collapse of man-made structures during the violent shaking of the ground. Accordingly, the most effective way to mitigate the damage of earthquakes from an engineering standpoint is to design and construct structures capable of withstanding strong ground motions.

Interpreting Recorded Ground Motions

Most elastic waves recorded close to an extended fault source are complicated and difficult to interpret uniquely. Understanding such near-source motion can be viewed as a three-part problem. The first part stems from the generation of elastic waves by the slipping fault as the moving rupture sweeps out an area of slip along the fault plane within a given time. The pattern of waves produced is dependent on several parameters, such as fault dimension and rupture velocity. Elastic waves of various types radiate from the vicinity of the moving rupture in all directions. The geometry and frictional properties of the fault critically affect the pattern of radiation from it.

The second part of the problem concerns the passage of the waves through the intervening rocks to the site and the effect of geologic conditions. The third part involves the conditions at the recording site itself, such as topography and highly attenuating soils. All these questions must be considered when estimating likely earthquake effects at a site of any proposed structure.

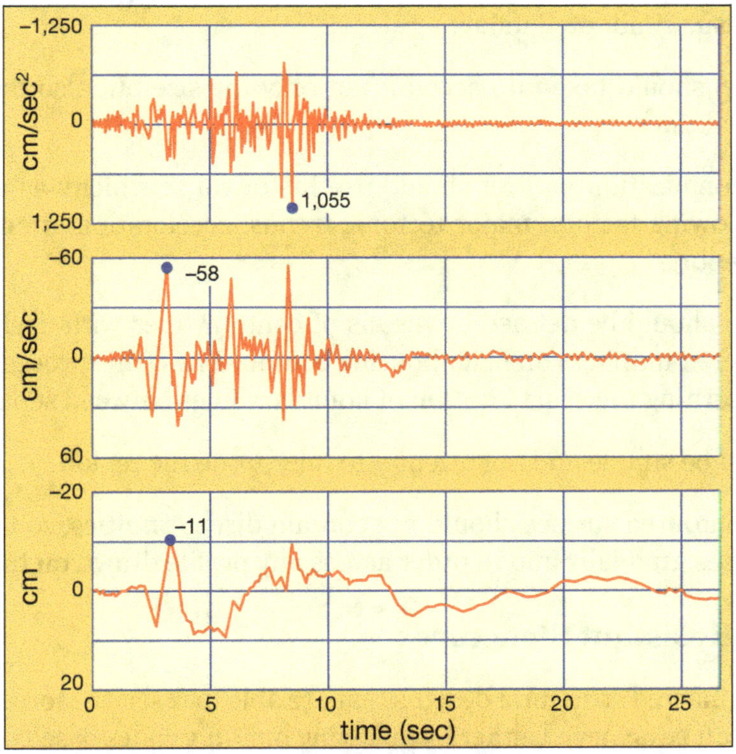

Recording of the San Fernando earthquake, near Pacoima Dam, California, 1971, showing (top) ground acceleration, (centre) velocity, and (bottom) displacement.

Experience has shown that the ground strong-motion recordings have a variable pattern in detail but predictable regular shapes in general (except in the case of strong multiple earthquakes). An example of actual shaking of the ground (acceleration, velocity, and displacement) recorded during an earthquake is given in the . In a strong horizontal shaking of the ground near the fault source, there is an initial segment of motion made up mainly of *P* waves, which frequently manifest themselves strongly in the vertical motion. This is followed by the onset of *S* waves, often associated with a longer-period pulse of ground velocity and displacement related to the near-site fault slip or fling. This pulse is often enhanced in the direction of the fault rupture and normal to it. After the *S* onset there is shaking that consists of a mixture of *S* and *P* waves, but the *S* motions become dominant as the duration increases. Later, in the horizontal component, surface waves dominate, mixed with some *S* body waves. Depending on the distance of the site from the fault and the structure of the intervening rocks and soils, surface waves are spread out into long trains.

Constructing Seismic Hazard Maps

In many regions, seismic expectancy maps or hazard maps are now available for planning purposes. The anticipated intensity of ground shaking is represented by a number called the peak acceleration or the peak velocity.

To avoid weaknesses found in earlier earthquake hazard maps, the following general principles are usually adopted today:

- The map should take into account not only the size but also the frequency of earthquakes.

The broad regionalization pattern should use historical seismicity as a database, including the following factors: major tectonic trends, acceleration attenuation curves, and intensity reports.

Regionalization should be defined by means of contour lines with design parameters referred to ordered numbers on neighbouring contour lines (this procedure minimizes sensitivity concerning the exact location of boundary lines between separate zones).

The map should be simple and not attempt to microzone the region.

The mapped contoured surface should not contain discontinuities, so that the level of hazard progresses gradually and in order across any profile drawn on the map.

Developing Resistant Structures

Developing engineered structural designs that are able to resist the forces generated by seismic waves can be achieved either by following building codes based on hazard maps or by appropriate methods of analysis. Many countries reserve theoretical structural analyses for the larger, more costly, or critical buildings to be constructed in the most seismically active regions, while simply requiring that ordinary structures conform to local building codes. Economic realities usually determine the goal, not of preventing all damage in all earthquakes but of minimizing damage in moderate, more common earthquakes and ensuring no major collapse at the strongest intensities. An essential part of what goes into engineering decisions on design and into the development and revision of earthquake-resistant design codes is therefore seismological, involving measurement of strong seismic waves, field studies of intensity and damage, and the probability of earthquake occurrence.

Earthquake risk can also be reduced by rapid post-earthquake response. Strong-motion accelerographs have been connected in some urban areas, such as Los Angeles, Tokyo, and Mexico City, to interactive computers. The recorded waves are correlated with seismic intensity scales and rapidly displayed graphically on regional maps via the World Wide Web.

Exploration of the Earth's Interior with Seismic Waves

Seismological Tomography

Seismological data on the Earth's deep structure come from several sources. These include *P* and *S* waves in earthquakes and nuclear explosions, the dispersion of surface waves from distant earthquakes, and vibrations of the whole Earth from large earthquakes.

One of the major aims of seismology is to infer the minimum set of properties of the Earth's interior that will explain recorded seismic wave trains in detail. Notwithstanding the tremendous progress made in the exploration of the Earth's deep structure during the first half of the 20th century, realization of this goal was severely limited until the 1960s because of the laborious effort required to evaluate theoretical models and to process the large amounts of earthquake data recorded. The application of high-speed computers with their enormous storage and rapid retrieval capabilities opened the way for major advances in both theoretical work and data handling.

Since the mid-1970s, researchers have studied realistic models of the Earth's structure that include continental and oceanic boundaries, mountains, and river valleys rather than simple structures such as those involving variation only with depth. In addition, various technical developments have benefited observational seismology. For example, the implications of seismic exploratory techniques developed by the petroleum industry (such as seismic reflection) have been recognized and the procedures adopted. Equally significant has been the application of three-dimensional imaging methods to the exploration of the Earth's deep structure. This has been made possible by the development of very fast microprocessors and computers with peripheral display equipment.

The major method for determining the structure of the Earth's deep interior is the detailed analysis of seismograms of seismic waves. (Such earthquake readings also provide estimates of wave velocities, density, and elastic and inelastic parameters in the

Earth.) The primary procedure is to measure the travel times of various wave types, such as *P* and *S*, from their source to the recording seismograph. First, however, identification of each wave type with its ray path through the Earth must be made.

Seismic rays for many paths of *P* and *S* waves leaving the earthquake focus *F* are shown in the figure. Rays corresponding to waves that have been reflected at the Earth's outer surface (or possibly at one of the interior discontinuity surfaces) are denoted as *PP*, *PS*, *SP*, *PSS*, and so on. For example, *PS* corresponds to a wave that is of *P* type before surface reflection and of *S* type afterward. In addition, there are rays such as *pPP*, *sPP*, and *sPS*, the symbols *p* and *s* corresponding to an initial ascent to the outer surface as *P* or *S* waves, respectively, from a deep focus.

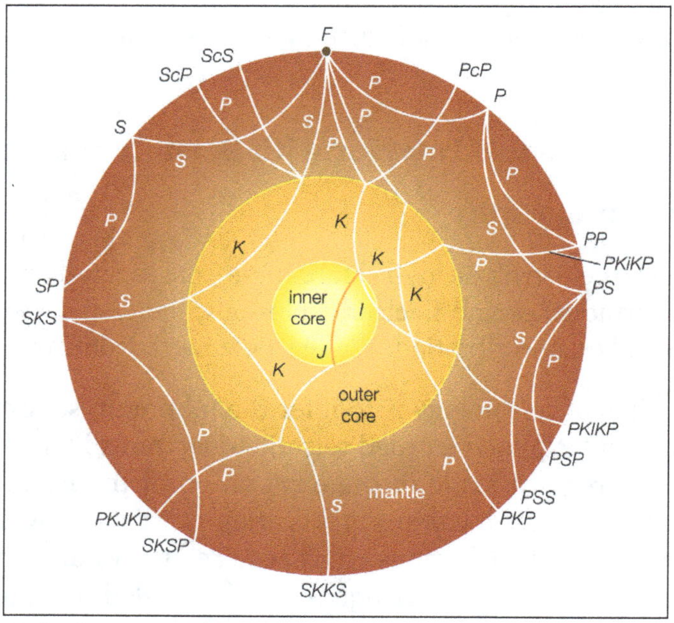

Seismic ray types in Earth's interior from an earthquake at *F*.

An especially important class of rays is associated with a discontinuity surface separating the central core of the Earth from the mantle at a depth of about 2,900 km (1,800 miles) below the outer surface. The symbol *c* is used to indicate an upward reflection at this discontinuity. Thus, if a *P* wave travels down from a focus to the discontinuity surface in question, the upward reflection into an *S* wave is recorded at an observing station as the ray *PcS* and similarly with *PcP*, *ScS*, and *ScP*. The symbol *K* is used to denote the part (of *P* type) of the path of a wave that passes through the liquid central core. Thus, the ray *SKS* corresponds to a wave that starts as an *S* wave, is refracted into the central core as a *P* wave, and is refracted back into the mantle, wherein it finally emerges as an *S* wave. Such rays as *SKKS* correspond to waves that have suffered an internal reflection at the boundary of the central core.

The discovery of the existence of an inner core in 1936 by the Danish seismologist Inge Lehmann made it necessary to introduce additional basic symbols. For paths of waves inside the central core, the symbols *i* and *I* are used analogously to *c* and *K* for the whole Earth; therefore, *i* indicates reflection upward at the boundary between the outer and inner portions of the central core, and *I* corresponds to the part (of *P* type) of the path of a wave that lies inside the inner portion. Thus, for instance, discrimination needs to be made between the rays *PKP*, *PKiKP*, and *PKIKP*. The first of these corresponds to a wave that has entered the outer part of the central core but has not reached the inner core, the second to one that has been reflected upward at the inner core boundary, and the third to one that has penetrated into the inner portion.

By combining the symbols *p*, *s*, *P*, *S*, *c*, *K*, *i*, and *I* in various ways, notation is developed for all the main rays associated with body earthquake waves. The symbol *J* has been introduced to correspond to *S* waves in the inner core, should evidence ever be found for such waves.

Finally, the use of times of travel along rays to infer hidden structure is analogous to the use of X-rays in medical tomography. The method involves reconstructing an image of internal anomalies from measurements made at the outer surface. Nowadays, hundreds of thousands of travel times of *P* and *S* waves are available in earthquake catalogs for the tomographic imaging of the Earth's interior and the mapping of internal structure.

Structure of the Earth's Interior

Studies with earthquake recordings have given a picture inside the Earth of a solid but layered and flow-patterned mantle about 2,900 km (1,800 miles) thick, which in places lies within 10 km (6 miles) of the surface under the oceans.

The thin surface rock layer surrounding the mantle is the crust, whose lower boundary is called the Mohorovičić discontinuity. In normal continental regions the crust is about 30 to 40 km thick; there is usually a superficial low-velocity sedimentary layer underlain by a zone in which seismic velocity increases with depth. Beneath this zone there is a layer in which *P*-wave velocities in some places fall from 6 to 5.6 km per second. The middle part of the crust is characterized by a heterogeneous zone with *P* velocities of nearly 6 to 6.3 km per second. The lowest layer of the crust (about 10 km thick) has significantly higher *P* velocities, ranging up to nearly 7 km per second.

In the deep ocean there is a sedimentary layer that is about 1 km thick. Underneath is the lower layer of the oceanic crust, which is about 4 km thick. This layer is inferred to consist of basalt that formed where extrusions of basaltic magma at oceanic ridges have been added to the upper part of lithospheric plates as they spread away from the ridge crests. This crustal layer cools as it moves away from the ridge crest, and its seismic velocities increase correspondingly.

Below the mantle lies a shell that is 2,255 km thick, which seismic waves show to have the properties of a liquid. At the very centre of the planet is a separate solid core with a radius of 1,216 km. Recent work with observed seismic waves has revealed three-dimensional structural details inside the Earth, especially in the crust and lithosphere, under the subduction zones, at the base of the mantle, and in the inner core. These regional variations are important in explaining the dynamic history of the planet.

Long-period Oscillations of the Globe

Sometimes earthquakes are large enough to cause the whole Earth to ring like a bell. The deepest tone of vibration of the planet is one with a period (the length of time between the arrival of successive crests in a wave train) of 54 minutes. Knowledge of these vibrations has come from a remarkable extension in the range of periods of ground movements that can be recorded by modern digital long-period seismographs that span the entire allowable spectrum of earthquake wave periods: from ordinary *P* waves with periods of tenths of seconds to vibrations with periods on the order of 12 and 24 hours such as those that occur in Earth tidal movements.

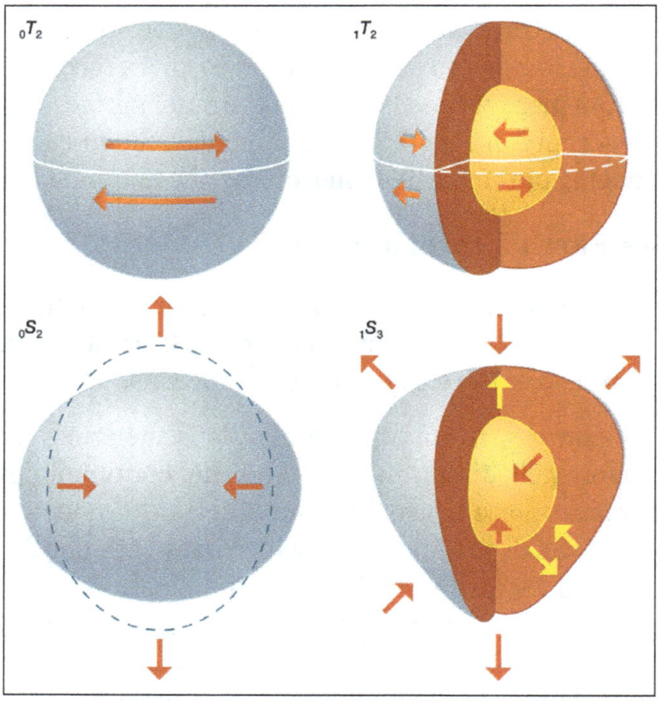

Displacements of the Earth in four types of free vibrations.

The measurements of vibrations of the whole Earth provide important information on the properties of the interior of the planet. It should be emphasized that these free vibrations are set up by the energy release of the earthquake source but continue for many hours and sometimes even days. For an elastic sphere such as the Earth, two types of vibrations are known to be possible. In one type, called *S* modes, or spheroidal vibrations, the motions of the elements of the sphere have components along the radius as well as along the tangent. In the second type, which are designated as *T* modes, or torsional vibrations, there is shear but no radial displacements. The nomenclature is *nSl* and *nTl*, where the letters *n* and *l* are related to the surfaces in the vibration at which there is zero motion. Four examples are illustrated in the figure. The subscript n gives a count of the number of internal zero-motion (nodal) surfaces, and l indicates the number of surface nodal lines.

Several hundred types of S and T vibrations have been identified and the associated periods measured. The amplitudes of the ground motion in the vibrations have been determined for particular earthquakes, and, more important, the attenuation of each component vibration has been measured. The dimensionless measure of this decay constant is called the quality factor Q. The greater the value of Q, the less the wave or vibration damping. Typically, for ${}_0S_{10}$ and ${}_0T_{10}$, the Q values are about 250.

The rate of decay of the vibrations of the whole Earth with the passage of time can be seen in the figure, where they appear superimposed for 20 hours of the 12-hour tidal deformations of the Earth. At the bottom of the figure these vibrations have been split up into a series of peaks, each with a definite frequency, similar to that of the spectrum of light. Such a spectrum indicates the relative amplitude of each harmonic present in the free oscillations. If the physical properties of the Earth's interior were known, all these individual peaks could be calculated directly. Instead, the internal structure must be estimated from the observed peaks.

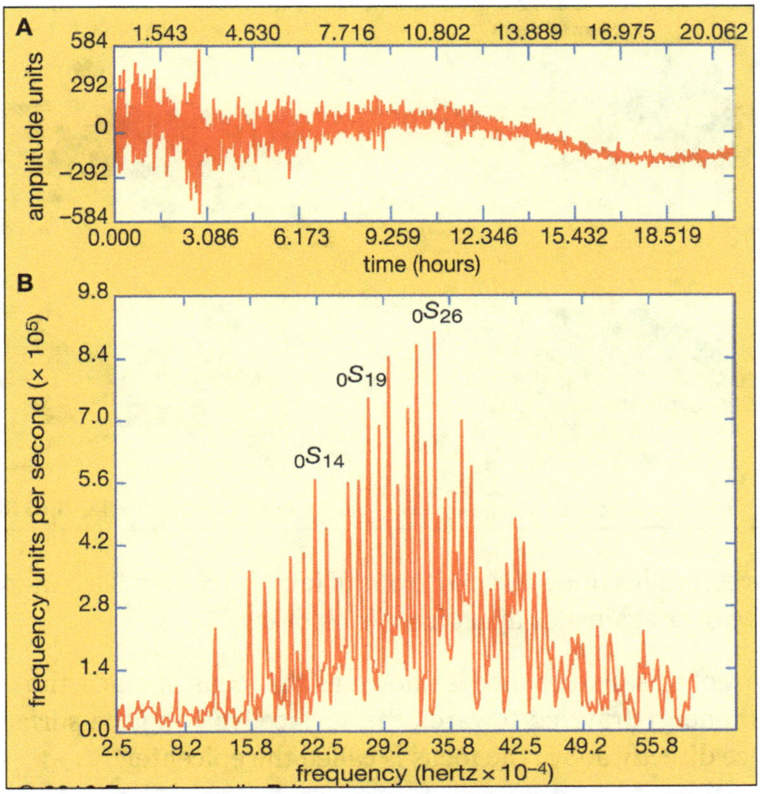

(A) Recorded ground motion for 20 hours at Whiskeytown, California, in the large Indonesian earthquake, 1977. (B) Frequency spectrum of the Earth's oscillations from that record.

Recent research has shown that observations of long-period oscillations of the Earth discriminate fairly finely between different Earth models. In applying the observations to improve the resolution and precision of such representations of the planet's internal structure, a considerable number of Earth models are set up, and all the periods of their free oscillations are computed and checked against the observations. Models can then be successively eliminated until only a small range remains. In practice, the work starts with existing models; efforts are made to amend them by sequential steps until full compatibility with the observations is achieved, within the uncertainties of the observations. Even so, the resulting computed Earth structure is not a unique solution to the problem.

Types of Earthquake

Tectonic Earthquake

The earth crust is made up of unevenly shaped slab of rocks called tectonic plates. The energy stored here causes the tectonic plates to move towards away or push against each other. With time the stored energy and the movement of the tectonic plates build up the enormous pressure within the region between two plates.

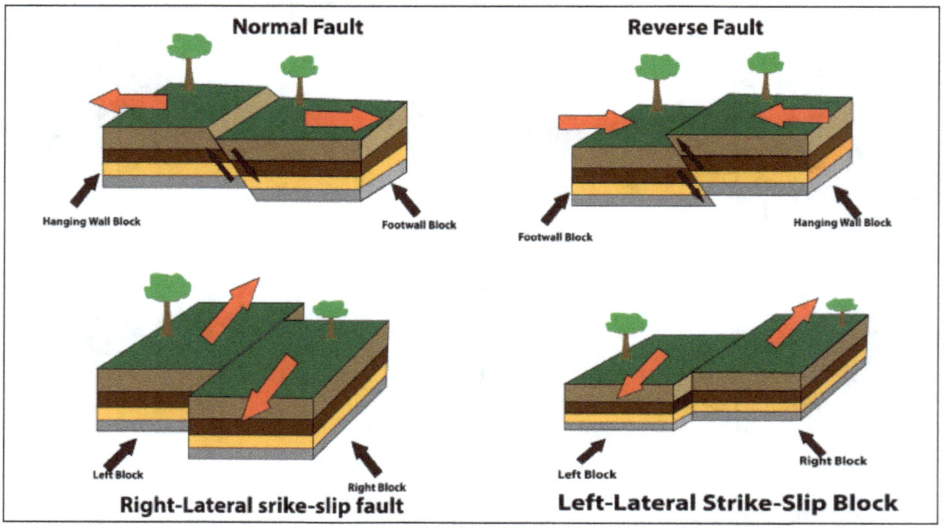

That intense pressure becomes the cause of the fault line(Like San Andrea Fault line), and plates move over against or apart from each other.

The center point of this disturbance is known as the focus of an earthquake. From the focus, waves of energy traveled towards the surface, shaking the surface within. The point the surface directly above the focus is called the epicenter.

From the epicenter of an earthquake, the energy waves traveled in a different direction on the earth surface causes vigorous movement on the surface of the earth which is known as an earthquake. A powerful earthquake can destroy an entire city.

Volcanic Earthquake

- Earthquake-related to the volcanic activity are called a volcanic earthquake.
- The magnitude of these quakes is usually weak.

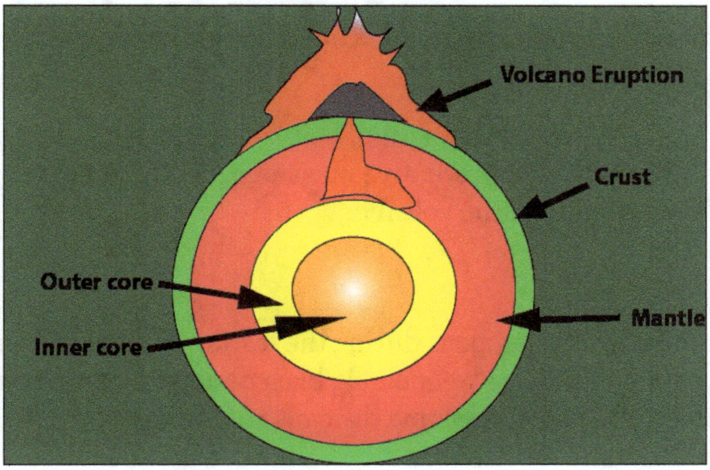

There are two types of volcanic earthquakes, which are:

- Volcano-tectonic earthquake.
- Long-period earthquake.

Volcano-tectonic Earthquake

- The tremors happen due to injection or withdrawal of Magma between the stressed rocks is called a volcano-tectonic earthquake.
- The movements of the molten magma cause mostly volcanic earthquake directly underneath a volcano.
- In these types of the earthquake, magma exerts the pressure on the tectonic plates until this magma breaks the rocks.
- During these cracks, tremors occur.
- These earthquakes are so weak that only can be measured by sensitive instruments.
- After the breakdown of rocks, magma starts flowing towards the surface of the earth.
- After that, rocks have to fill up space where magma is no longer available. During filling the space, tremors happen of very low magnitude.

Long-period Earthquake

- A long period earthquake occurs while the injection of magma into the surrounding rocks. These all happen due to the pressure changes among the layers of the earth.
- These types of volcanic activity indicate the eruption of the volcano in the near future.
- Seismograph records these reading, and before the actual eruption of the volcano, we can easily take preventive measures. So that local people can become aware of the eruption of the volcano.

Explosive Earthquake

Explosive earthquake mostly happens during the testing of nuclear weapons. We know that during detonation of nuclear weapon big blast occurs and a vast amount of energy releases. Sometimes these blasts become the cause of the earthquakes.

Collapse Earthquake

- These earthquakes are of weak magnitude earthquakes happen in the caverns and mines.
- Sometimes, underground blasts (Rock breaking) in the mines become the cause of the collapsing of mines and collapsing of mines produces seismic waves.
- The P and S- waves (Seismic waves) produced during the explosion of rocks on or under the surface of the earth cause this type of earthquakes.

EARTHQUAKE PREDICTION

Prediction is concerned with forecasting the occurrence of an earthquake of a particular intensity over a specific locality within a specific time limit. Normally prediction is of three types viz. long, medium and short range prediction.

While long range prediction is concerned with forecasting the occurrence of an earthquake a number of years in advance, medium term prediction is to be done a few months to a year or so and the short term prediction implies forecast ranging from a few hours to some days in advance.

Medium and short range predictions are very useful because they can help in saving the largest population from disaster in terms of life and property. Scientists believe that it is possible to predict major earthquakes by monitoring the seismicity caused by natural earthquakes, mining blasts, nuclear tests, etc.

However, no flawless technique has been developed to predict the earthquakes till date. Most of the methods and models are beyond the scope of the present work and only a few simple methods and models will be discussed here.

Unusual Animal Behavior

It is a well established fact that animals are endowed with certain sensory perceptions denied to human beings. Some of the animals have much better power of sniffing, hearing, seeing and sensing than the human beings. The unusual behaviour of animals prior to earthquakes received wide publicity after the Haichang earthquake in Liaoning province of China, in February 4, 1975 was successfully predicted.

Although fluctuations in water levels and radon content in water were given due consideration, behaviour of animals was not overlooked in the process of earthquake prediction. On the morning of February 4, 1975, a moderate forestock hit the city of Haichang and by 2 p.m. a general alert was proclaimed.

Within six hours, the area was rocked by a devastating earthquake of 7.3 magnitude but almost all the one lakh residents were saved. Chinese are considered to be pioneers in recognising the unusual behaviour of animals preceding a quake as an important indicator to predict an impending earthquake, particularly since the accurate prediction of Haichang quake of 1975.

In fact, national war against earthquakes was launched in 1966 with an effective slogan, "Rather a thousand days with no earthquake than one day with no precaution." Chinese report was presented at the Intergovernmental meeting convened at UNESCO, Paris in February 1976. This stimulated considerable interest among scientists.

However, it should be mentioned that abnormal behaviour of animals prior to a devastating earthquake was noticed earlier also in different parts of the world. In Japan large number of rats were seen every day in a restaurant in Nagoya City, which suddenly disappeared on the evening prior to Nobi earthquake of 1891.

Similar observations about rats were reported at two earlier occasions i.e., Kanto earthquake of 1923 and Sankriku earthquake of 1933. In China, unusual behaviour of rats was reported before 1966 Hsingtai earthquake in Hopei Province (300 km from Beijing).

In 1835 dogs escaped from the city of Talcahuano in Chile before the earthquake struck the city. Flocks of birds flew inland before the Chilean earthquakes of 1822 and 1835. Monkeys became restless a few hours before the Managua earthquake of 1972 in Nicaragua.

In summer of 1969, just before the Bahai quake, the custodians of Tientsin Zoo had observed that swans suddenly scrambled out of water and stayed away, a Manchurian tiger stopped pacing, a Tibetan yak collapsed, pandas held their heads in paws and moaned; and turtles were restless.

Hens and cocks were reported restless about an hour's before in 1896 Ryakya earthquake in Japan. In Yugoslavia, birds in zoo started crying before 1963 earthquake. Deer gathered and cats disappeared from villages in northern Italy two or three hours before damaging earthquake of 1976.

Just before the earthquake which occurred in 1906 along the San Andreas Fault, horses whined and cows stampeded. In other cases cows about to be milked became restless before the shock. Bellowing of the cattle at the time of shock was very commonly reported. Howling by dogs was reported during the night preceding the earthquake.

Abnormal behaviour just before an earthquake has also been noticed among animals who live underground, like snakes, insects and worms, and those living in water (fishes). Abundant fishes were caught in just before the 1896 earthquake in north western coast of Japan and the Tango earthquake of 1927. However, in Kanto earthquake fishes were reported to have disappeared.

Just before the Edo earthquake, many grass snakes were reported to have come out of the ground near the epicentral areas even though it was severe cold winter. Very unusual behaviour of dogs was reported just before the Turkey earthquake. Barry Ralleigh of the U.S. Geological Survey noticed that the horses were fidget just before the earthquake of 28 November 1974 in Hollister.

Extensive research is being carried on all over the world about the unusual behaviour of animals with respect to prediction of the earthquake. China and Japan are fore-runners in this regard. The USA has also shown keen interest in unusual behaviour of animals as a useful indicator of earthquake prediction.

The Stanford Research Institute, California, under the 'Project Earthquake Watch' has a network along the San Andreas Fault. This group keeps a watch on the behaviour of about 70 animal species. Dr. B.G. Deshpande has compiled a list of 87 animals which have been watched all over the world and whose behaviour might sense as an advance indicator of impending quake. Some of these which may be easily observed by the city dwellers are; cockroaches, crows, dogs, donkeys, ducks, fowls, frogs, geese, goats, horses, mice, monkeys, pigs, pigeons, rats, sheep, squirrels, swans and snakes.

The Group of Earthquake Research Institutes of Biophysics, China has arrived at the following conclusions after an extensive survey of animal behaviour before a strong earthquake:

- Most animals show increased restlessness before an earthquake.
- The precursor time varies from a few minutes to several days, with increased restlessness at 11 hours which becomes still more marked about 2 to 3 hours before the earthquake. In general precursor times of various animals are mostly within 24 hours before the earthquake.

- These observations have been noticed predominantly in high intensity or epicentral region close to active faults.
- Abnormal behaviour of the animals is observed during earthquakes of magnitude 5 or more.
- More intensive response can be noticed with the increase of intensity of earthquakes.

Hydrochemical Precursors

Chemical composition of underground water was observed on a regular basis in seismically active regions of Tadzhik and Uzbekistan. These observations yielded following results:

- Concentration levels of dissolved minerals and gaseous components remained almost constant during seismically inactive period.
- An appreciable increase in concentration of dissolved minerals was noticed 2 to 8 days before an earthquake. Variations in level of underground water, the pressure of artesian water, the discharge of water sources and temperature of underground water were also noticed during this period. These variations are large in the event of a strong earthquake.
- After the earthquake, anomalies in concentrations of the gaseous and mineral components disappear.

Temperature Change

There seems to be some relation between temperature and earthquakes. A considerable rise of temperature by 10 °C and 15 °C was reported before earthquakes in Lunglin in China and Przhevalsk in Russia. The epicentral distances of these earthquakes where observations were taken in hot spring/well were 10 and 30 km and precursory periods were 42 and 72 days respectively.

Water Level

There are drastic changes in water level in several wells just before a major earthquake. There was a fall in water level a few days before the Nankai earthquake in Japan. Rise of water level by 3 and 15 cm was reported before Lunglin (China) and Przhevalsk (Russia) earthquakes.

Similarly, water level rose by 3 cm a few hours before the earthquake in Meckering in Australia. In China rise of water level in wells was observed before earthquakes of Haicheng, Tangshan, Liu- quiao and Shanyin.

Experiments in water level variations have been conducted in Kurile Islands to predict the earthquakes of 4 and more on Richter scale. For this purpose wells upto 410-670 metres depth at epicentral distances upto 700 km are used. This is an effective technique for observing the deformation of the earth's crust. The model on which the forecasts of earthquakes is based shows that 3 to 10 days before an earthquake, the water level begins to fall. After a short period, it starts rising when the earthquake strikes.

Radon Gas

Radon is a radioactive gas which is discharged from rock masses prior to earthquake. It is dissolved in the well water and its concentration in the water increases. Such an increase was reported in Tashkent in 1972 where increase in concentration varying from 15 to 200 per cent was noticed about 3 to 13 days prior to an earthquake.

In China, 50% and 70% increase in radon concentration was reported 18 and 6 days respectively before the Tangshan and Luhuo at Langfang and Guzan stations which were located 130 and 200 km epicentral distances for two cases. In 1995, a correlation in radon anomalies at four sites in Kangra and one site in Amritsar with the time of occurrence of Uttarkashi earthquake was reported.

Oil Wells

Large scale fluctuations of oil flow from oil wells prior to earthquakes were reported in Israel, northern Caucasus (Europe) and China. These earthquakes which occurred in 1969, 1971 and 1972 gave rise to increased flow of oil before their occurrence. It has been suggested that when the tectonic stress accumulates to a certain level, the pore pressure within a deep oil bearing strata reach its breaking strength causing oil to sprout along the oil wells.

Theory of Seismic Gap

Seismicity gap is a region where earthquake activity is less compared with its neighbourhood along plate boundaries. Soviet seismologist S.A. Fedotov studied the seismic record of 12 large earthquakes which rocked northern Japan between 1904 and 1963. By plotting the size of each tremor- struck area, he found that each quake segment abutted the next contiguous one without overlapping, as if each deep seated crack had been shut off by a barrier at the ends of the fracture zone.

Each large earthquake was in a segment that was quiet for the last 39 years or so. Fedotov predicted that those segments which were quiet for some time will be hit by earthquake sooner or later. Three of these blocks in Kurile Island were struck where according to Fedotov an earthquake was due. Thus evolved the theory of seismic gap in earthquake prediction.

Based on this theory Dr. Kiyo Mogi of Tokyo succeeded in predicting a few earthquakes in Japan. Three geophysicists—Masakazu Ohtake, Tosimatu Matumoto and Gary V. Latham—working at Taxas University's Marine Science Institute had predicted a major earthquake in southern Mexico around the town of Puerto Angel based on the theory of seismic gap.

On 29 November, 1978, a severe earthquake measuring 7.9 on the Richter scale with an epicentre within a kilometre of the predicted site struck the area. A seismic gap predicted quake also occurred along the San Andreas Fault.

Foreshocks

Generally major earthquakes are preceded by minor shocks known as foreshocks. These foreshocks provide valuable dues to the occurrence of a strong earthquake. Some of the earthquakes have been successfully predicted on the basis of study of foreshocks. In addition to unusual behaviour of animals, the Haichang earthquake in China was predicted by studying the increased seismicity from December 1974 to February 1975.

The Oaxaca, Mexico earthquake of November 1978 was also successfully predicted on the basis of foreshock observations. Foreshocks have been detected a few days to a month in advance with the help of closely located seismic stations in Himachal Pradesh for several earthquakes like Anantnag, Dharmasala, Kashmir, Kinnaur and a few others. Uttarkashi earthquake of October 20, 1991 was preceded by foreshocks on October 15 and 16 with magnitude larger than 3.5 on Richter scale.

The most recent Bhuj earthquake of January 26, 2001 was also proceeded by foreshocks in December 2000. But there are some other earthquakes which are proceeded by foreshocks. Therefore, this is not a flawless method and has to be supplemented by other methods of earthquake prediction.

Changes in Seismic Wave Velocity

We know that P, S, and L waves originate from the focus of an earthquake. P and S are called body waves because they travel through the body of the earth, while L waves are known as surface waves because they move along the upper crust of the earth. P waves are faster than the S waves and reach seismographs first.

The time lag between the arrival of P and S waves is called lead time. Russian seismologists found that this lead time began to decrease significantly for days, weeks and even months before the earthquake. But just before the quake hit the area the lead time was back to normal. A longer period of abnormality in wave velocity presaged a larger quake.

Taking the cue from the Russians, Lynn Sykes, Scholz and Aggarwal conducted laboratory, experiments on rock samples in 1973. These experiments showed abnormal change of ratio of velocities of P waves and S waves before the earthquake.

This ratio is expressed as V_p/V_s. The duration of V_p/V_s anomaly depends upon the fault or dimensions of the aftershock area. After the Garm region of the former USSR, V_p/V_s anomalies were noticed in Blue Mountain Lake earthquake in the USA in 1973. The velocity anomaly period for this earthquake was about 5 days and the decrease in velocity was about 12 per cent.

Similar decrease in velocity ratio was reported before the damaging Haichang, Songpan-Perigwu and bungling earthquakes in China. In Japan, 7 to 40% decrease in the velocity ratio ranging from 50 to 700 days before the main earthquakes were recorded. In Tehran 14% decrease in velocity was reported 1 to 3 days before three earthquakes in 1974.

Immediately after the Gujarat earthquake of 2001, the Survey of India mooted a network of 300 permanent Geographical Positioning System (GPS) stations all over the country to monitor earth movements round the clock—which help in predicting earthquakes.

If the GPS systems are located along the known active faults, it is possible to monitor movements of active faults or breaks in the earth's crust. Though no precise prediction can be made about the location and magnitude of an earthquake, minor movements are an indication of an impending earthquake because it reflects the force coming from below the crust.

EARTHQUAKES AND STRUCTURAL DAMAGES

Earthquakes are the most destructive natural hazards throughout human history. Hundreds of thousand people lost their lives and loss of billions of dollars' properties occurred in these disasters. Occurred medium or high-intensity magnitude earthquakes in last twenty years showed that these loses continue. For reinforced concrete (R/C) buildings, inappropriate design such as soft and weak stories, strong beam–weak column, short column, hammering, unconfined gable wall and in-plane/out-of-plane movement of the walls causes damages. These are the main reasons. In addition to this, low quality of structural materials, poor workmanship, lack of engineering services, and construction with insufficient detailing of the structural elements are the another reasons of damages. Main reasons of masonry building damages in terms of design faults can be shown as heavy earthen roofs, inappropriate detailing of wall to wall connection and wall to roof connection, absence of bond beams, large openings. However, construction of buildings by using local materials with poor workmanship on the base of traditional rules is the other reason of failures for these buildings.

Earthquake Characteristics

Structure of the Earth

The earth consists of layers which have different properties. The outer layer of the earth is called as "crust." The thickness of this layer is between 35 and 70 km for continents, and this thickness varies between 5 and 10 km thickness for ocean floor. Mantle layer, existed under the crust, is divided as lower mantle and outer mantle. This layer is approximately 2900 km thickness. Convection current occurred in the mantle causes plate tectonics in the crust. Core is the innermost layer and divides into two parts as fluid outer core and solid inner core. The outer layer is 2300 km, and inner layer is 1200 km thickness. The internal structure of the earth is shown in the figure.

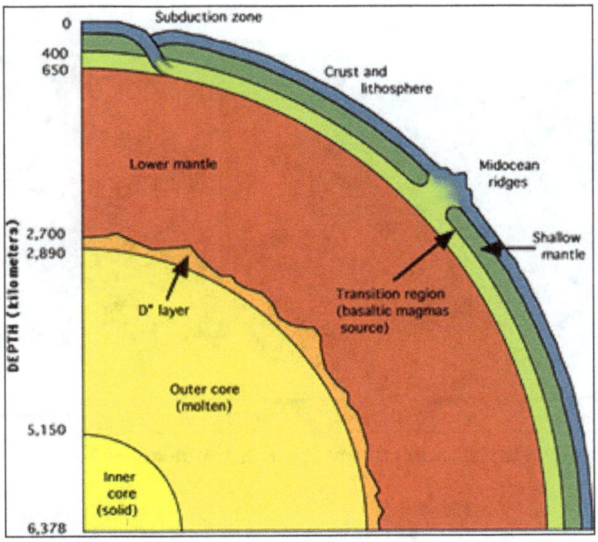

The internal structure of the earth.

Plate Tectonics

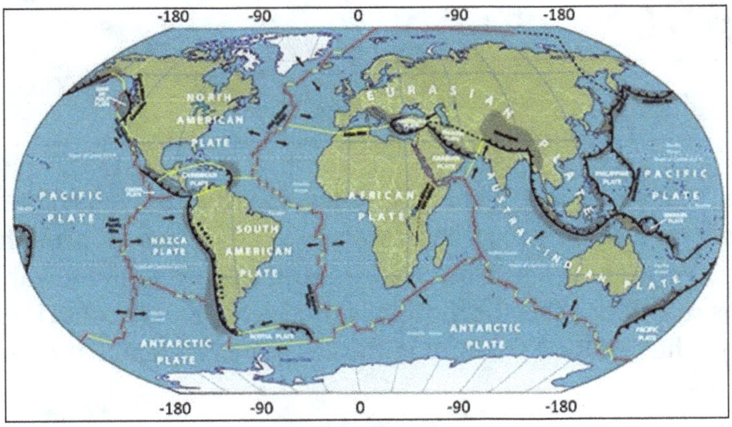

Tectonic plates.

Plate tectonic deals with movement and strain of earth crust. According to the state of the art of Plate tectonic, the earthquake occurs in some parts of the plate and these parts act relative to each other. Pressure shift arose from these action and cooling stages in mantle causes stresses in the earth crust. When the increased stresses reached to bearing capacity of the crust on faults, this event causes sliding (breakthrough). Sliding movement spreads outward starting from hypocenter. Strain energy, which cumulated for a long time, discharges with sliding and causes earthquake shaking. Propagation of wave from hypocenter that results surface sliding is perceived as earthquake. Figures below show the tectonic plates and worldwide earthquake distribution, respectively.

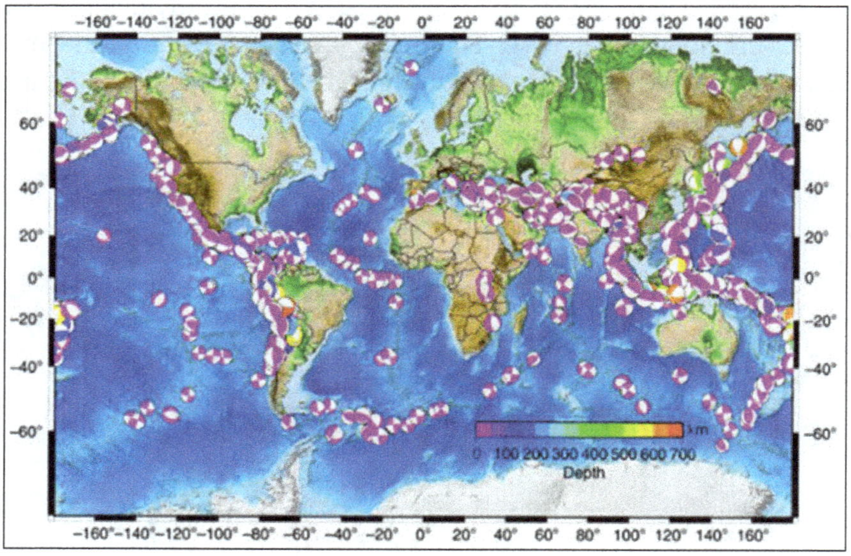

Geographical distribution of the 1700 earthquakes on the worldwide.

Faults

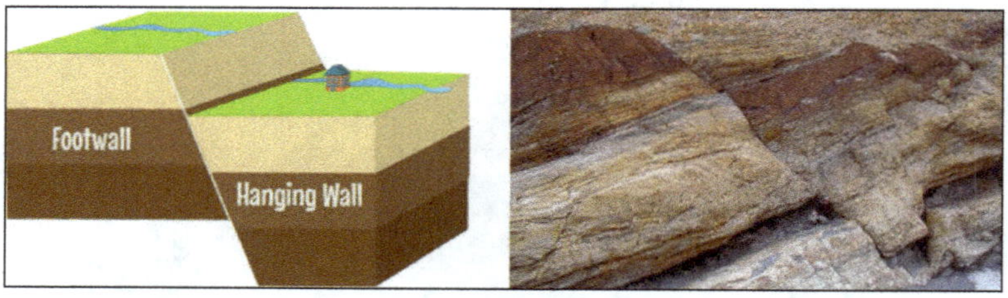

Normal fault graphical presentation and mechanism.

When two plates move with respect to each other, elastic strain energy is accumulated in results of tectonic processes. Thus, these two plates are released through the rupture of the interface zone. The shapeless blocks show immediate reaction towards equilibrium. As a result of this reaction, a seismic motion is produced. This process is called as "elastic rebound" theory. The resulting fracture in the crust of earthquake is defined as "fault".

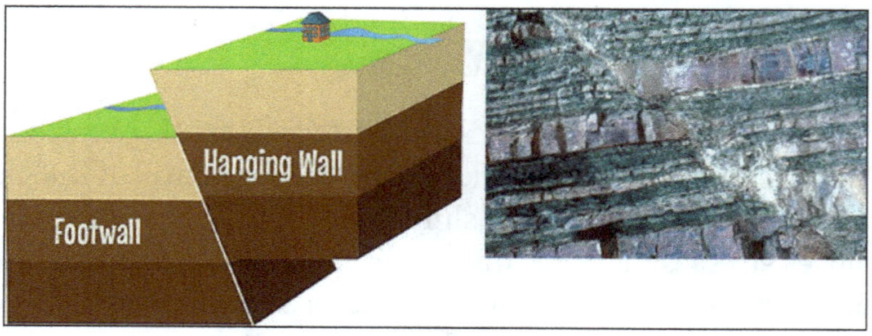

Reverse fault graphical presentation and mechanism.

Strip slip fault graphical presentation and mechanism.

Seismic Waves

Body Wave

Seismic activity that results in earthquake generates two types of seismic waves: body and surface. Body waves move through the interior layers of earth's. Body waves include primary waves (known as P-waves) and secondary waves (also called as S-waves). P-waves generate sequential push (or compression) and pull (or tension) in soil. P waves have relatively little damage potential. On the contrary, S-wave propagates horizontal and vertical motion. S-waves produce shear stresses in the soil along their paths.

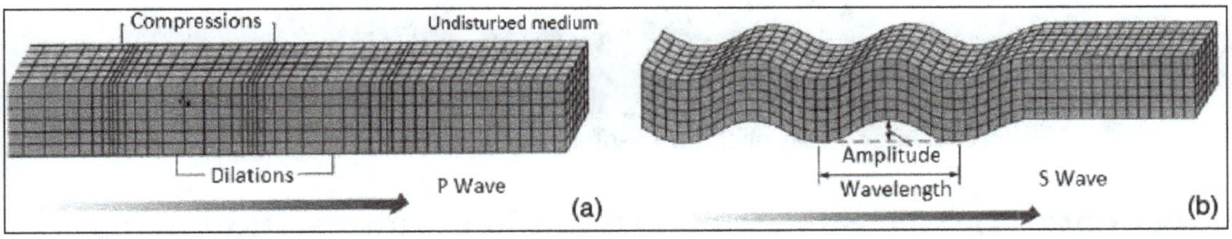

(a) Primary (P) wave and (b) Secondary (S) wave.

Surface Waves

Surface waves include Love (L) and Rayleigh (R) waves that propagate through the outer layers of the crust. These waves are generated by body waves move through parallel to the ground surface and various underpass the layer boundaries. These waves cause large displacements. These types of waves take various forms at a further distance away from the earthquake source. Surface waves are occurred during shallow earthquakes; on the other hand, body waves take place at all depths. Surface waves cause serious damage to structures due to their long duration.

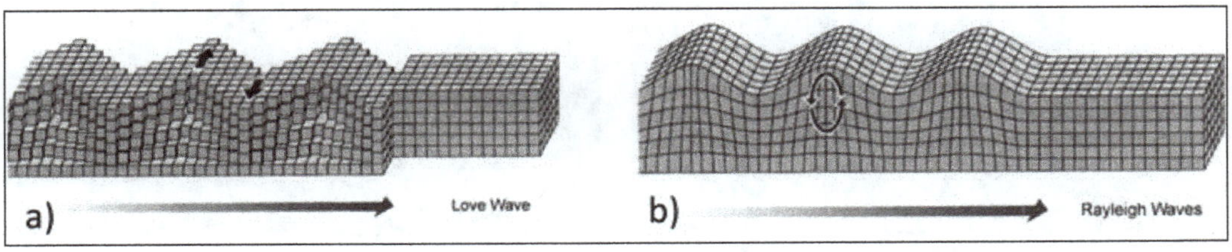

(a) Love (L) wave and (b) Rayleigh (R) wave.

Response of Reinforced Concrete (R/C) Structures

Soft and Weak Storey Mechanism

In some R/C buildings, especially at the ground floor, walls may not be continuous along to height of building for architectural, functional, and commercial reasons. While ground floor generally encloses with glass window instead of brick infill walls, partition walls are constructed above from this storey for separating rooms for the residential usage. This situation causes brittle failures at the end of the columns. In mid-rise reinforced concrete buildings, the most common failure mode is soft-storey mechanism, particularly at the first storey. Failures can be concentrated at any story called as weak storey in which the lateral strength changes suddenly between adjacent stories due to lack of or removing of partition walls or decreasing of cross section of columns. Thus, during an earthquake, partial and total collapses occur in these storeys.

Unexpected inter-storey drift due to soft storey during the Van earthquake.

Inadequate Transverse Reinforcement in Columns and Beams

Shear forces increase during an earthquake especially at columns and beam–column joints. Consequently, special attention should be paid to construction and design of beam–column joints and columns. Seismic design requires increasing of ductility of structures for performance-based design approach. In particular, columns of buildings can be having insufficient transverse reinforcement in the plastic hinge region. Therefore, structural elements which have such details show low performance against to dynamic loads and lost their shear and axial load carrying capacity.

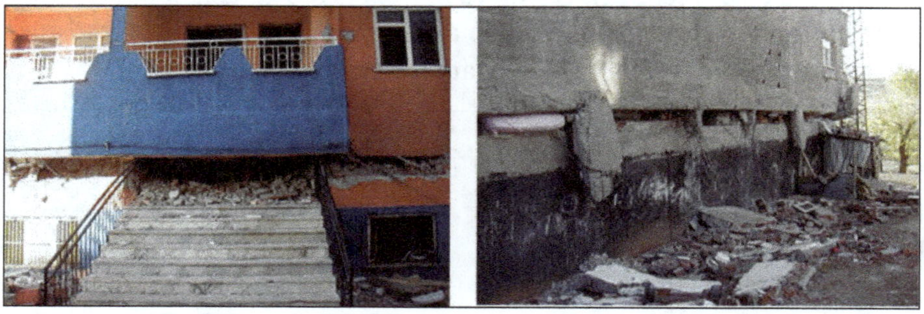

Weak storey mechanism during the Bingöl earthquake.

Damaged structure due to inadequate spacing between shear reinforcements during (a) Van earthquake and (b) Bingöl earthquake.

Short Column

Short column damages during the 2003 Bingöl earthquake.

This type of mechanism can be developed due to structural adjustments and/or to continuous openings at the top of infill walls between columns. Lateral forces that occurred by an earthquake are carried by columns and shear walls. Length of column is an important factor for dissipation of these loads. When the length of column decreases, the column becomes stiffer and brittle than the other columns and this column attracts more shear forces. Thus, shear failure which is a critical type of concrete column damage occurs at these columns.

Inadequate Gaps between Adjacent Buildings

Collapse of adjacent buildings during the Bingöl earthquake.

Buildings are sometimes constructed adjacent because of the lack of building lots. In this layout plan, one or two faces of two buildings are in contact to each other.

Consequently, the buildings that have not adequate gaps pound to each other during the earthquake. If the floors of the buildings are not at the same level, pounding effect of the buildings becomes more dangerous.

Strong Beam–weak Column

Failure of a building due to strong beam–weak column effect during the Van earthquake.

Deep and rigid beams are used with flexible columns in type of buildings. Therefore, these beams resist more moments, occurred by dynamic loads, than weak columns. In such a design during an earthquake while deep and rigid beams show elastic behavior, shear failure or compression crushing causes plastic hinges at flexible columns.

Failures of Gable Walls

The most common failure mode at gable walls is out-of-plane collapse in the earthquakes. Although failures of gable walls are not structural damages, these damages may be cause loss of lives and properties. Stability problems and large unsupported wall lengths cause damages at these walls.

Failure of gabble walls on top of the building during the Van earthquake.

Poor Concrete Quality and Corrosion

The other main reasons of damages are low concrete strength and workmanship. Concrete quality is an important factor for building performance against to earthquakes. Handmade concrete is used to without using vibrator in construction of old buildings. Thus, homogeny mixing was not obtained and expected compressive strength was not provided in these buildings. In addition to this, using of aggregates which have improper granulometry, corrosion which decreases reinforcement bar area, and using of smooth steel reinforcement effected strength of concrete.

Failure of column due to poor concrete quality during;
(a) Van earthquake and (b) Bingöl earthquake.

In-plane/Out-of-plane Effect

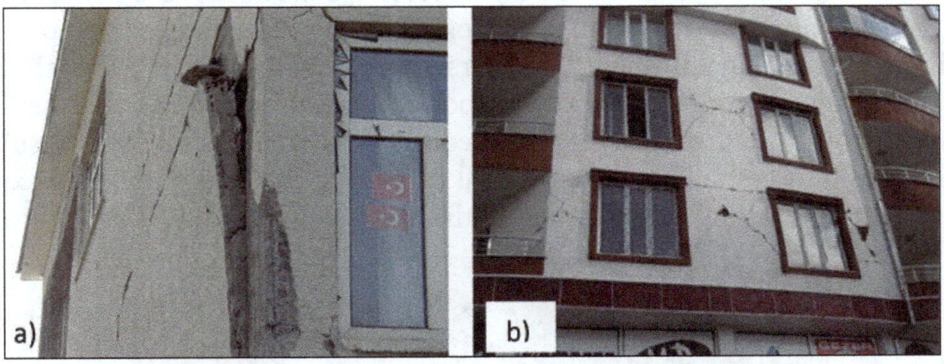

(a) Detachment of infill wall during the Bingöl earthquake and
(b) In-plane damage of during the Van earthquake.

One of the most important reasons of life and economic loss during the earthquake is combined effect of in-plane and out-of-plane movement of the wall. In-plane and out-of-plane interaction is very complicated and should be analyzed well for this phenomena. For low-rise and mid-rise unreinforced masonry (URM) infilled R/C frames, ground story infill walls are expected to be damaged firstly, because they are subjected to the highest in-plane demands. However, under the effect of bidirectional loading, where the two components of a ground motion are equally significant, infill walls of the upper stories may fail under the combination of in-plane and out-of-plane effects. The in-plane demand reduces at the upper stories, while that of out-of-plane forces increases due to the increase of accelerations. To prevent this problem, in-plane carrying capacity of the wall should increase and out-of-plane ductility should increase with possible and applicable developments like bed-joint reinforcements and wire mesh. These listed applications will prevent detachment of infill wall from reinforced concrete elements and will increase the stiffness of the total structural system.

Response of Masonry Structures

Earthen Roof Damages

One- and two-story masonry buildings are common in the rural areas because they require easy workmanship. These buildings are constructed with thick adobe and stone walls and mostly vulnerable to strong ground motions. But masonry buildings are preferred because of the advantages such as thermal properties and simple construction by using local materials. These structures are constructed with traditional techniques using locally available materials. Nearly no engineering services are used in these buildings. Adobe blocks are produced from local material contains mixed soil with straw and leave dried under the sun. These units are easily broken into small pieces as they have very low strength. As adobe blocks have low strength, walls of masonry buildings are thick and massive.

Generally, earthen roofs are constructed over wooden logs supported by two main walls of the buildings to provide thermal and water insulation. Also, heavy earthen roofs are constructed by the residents because a thick roof keeps the house cool or warm according to the seasons. However, these roofs lose their effectiveness because of weather conditions, such as rain and snow. Therefore, the residents place a new earthen cover on top of the existing roof to repair these roofs. Thus, the weight and thickness of the roof increase over time. As a result of these heavy roofs, the structures are subjected to larger inertia forces during the earthquakes. During the horizontal displacements, these heavy roofs are very vulnerable since they can easily slide over the walls.

Damage to adobe structures due to heavy earthen roof during the Van earthquake.

Corner Damages

Corner damages are common in the adobe and masonry buildings. This type of mechanism generally occurs at wall-to-wall and wall-to-roof connections when subjected to out-of-plane displacements. During an earthquake, the stress concentrations increase at intersection of the walls. In this way, vertical or inclined cracks appear in the corners of masonry buildings. If bond beams are not used in the corners or connection, two walls are not properly anchored each other, intensity of the cracks increases and these cracks spread along the height of the wall. Similar cracking may have been observed at adobe buildings. Poor connections between adjacent walls and the absence of bond beams cause serious damages. In addition, there are no appropriate connections at the corner of the walls in damaged buildings. Due to reduced compression stress and increase seismic acceleration at upper stories, the common failures are seen at the corners of the roof level. When there is no slab with some in-plane rigidity at roof level, top corners are more sensitive to fail because of cantilever-like behavior.

Corner damage of briquette masonry buildings.

Out-of-plane Mechanism

Out-of-plane mechanism may appear from the combination of several deficiencies. A lack of bond beams, poor connections among the walls and the roofs, and large unsupported wall lengths cause the separation of walls and cause damage to occur via the of out-of-plane mechanism. Thus, the whole or the significant parts of the wall fall down during the earthquake. Wooden logs that bear the weight of the floor of the building are generally placed on load-bearing walls in only one direction. Thus, earthquake loads are transferred to perpendicular walls by wooden logs. Therefore, the walls that are not supported by the wooden logs may easily overturn to out-of-plane direction during the earthquake. This failure mechanism can be commonly observed in the earthquake region.

Out-of-plane mechanism of the briquette buildings.

In-plane Mechanism

Diagonal shear (a-b), horizontal (c) and stepped (d) failures wall.

The seismic performance of the masonry buildings relates to the in-plane stiffness of the walls. In-plane mechanism is generally observed in most of the masonry buildings that are affected by shear cracking. Earthquake loads increase the shear force. It can damage walls and their connections. These damages generally occur near openings, because most of the masonry buildings don't have sufficient and proper bond beams that distribute the lateral forces uniformly and enhance the lateral strength of the walls. During the earthquakes, excessive bending and shear can produce in-plane failures depending on the aspect ratio of the unreinforced masonry elements. In the areas struck by earthquake, three failure modes of the shear damages in masonry buildings are generally observed, namely diagonal shear failures that proceed through masonry units and mortar, sliding consisting of straight failure at the horizontal bed joints, and stepped failures from the head to bed or bed to head joints.

Disintegration of Stone Masonry Walls

Most of the masonry buildings are constructed with thick stone walls. As the thickness of the walls is relatively large, these stone walls are constructed by using more than one stone along the thickness direction. In these walls, stones are placed in a random order. These walls have two exterior vertical wythes of large coarse stones. However, smaller rubble stones with mud mortar are used between two exterior layers. Using mud mortar instead of cement mortar causes insufficient adherence between the layers. Thus, the interior and exterior layers of the wall behaved independently and separated each other during the earthquake. Some reasons such as the quality of construction, poor workmanship, and the use of improper materials increase the intensity of the disintegration.

Disintegration in stone masonry buildings.

EARTHQUAKE ENGINEERING

Earthquake Engineering is a specialised field within Geotechnical Engineering. Earthquakes are simply a series of vibrations due movements below the Earth that have the potential to cause immense damage and destruction at the Earth's surface. Earthquake Engineering is the application of the study of earthquakes and the vibrations generated at the surface in order to protect the built environment and people from being harmed.

In areas that are significant risk of being exposed to earthquakes, a team of Earthquake Engineers are called in to prevent disaster from occurring. Importantly, they make significant use of design and construction techniques to ensure that buildings and infrastructure are able to withstand such considerable damage.

Simulation and failure mode analysis are important tools that help Earthquake Engineers to design structural elements that can withstand earthquake exposure without major structural damage.

GEOGRAPHICAL DISTRIBUTION OF EARTHQUAKES

Earthquakes can happen in any part of the world. But in the areas of faulting and folding or of crustal weakness, the frequency of earthquakes is more than anywhere else. The earthquakes are concentrated in two main belts.

- Circum-Pacific Earthquake Belt: This belt includes all the coastal areas around the vast pacific ocean. This belt extends as an isostatically sensitive zone through the coasts of Alaska, Aleutian Islands, Japan, Philippines, New Zealand, North and South America. This zone accounts for 68% of all earthquakes on the surface of the earth. The most talked about earthquake areas in this zone include Japan, Chile, California and Mexico.
- Mediterranean-Asia Earthquake Belt: This belt begins from Alps mountain range and passes through Turkey, Caucasus Range, Iran, Iraq, Himalayan mountains and Tibet to China. One of its branches passes through Mongolia and Lake Baikal and another branch extends to Myanmar. About 31% of world's earthquakes are located in this region.
- Other Areas: These include Northern Africa and Rift Valley areas of the Red Sea and the Dead Sea. In addition to these, the ocean ridges are also active earthquake zones.

References

- Earthquake-geology, science: britannica.com, Retrieved 21 May, 2020
- Earthquake, entry: newworldencyclopedia.org, Retrieved 11 April, 2020
- Earthquake-geology, science: britannica.com, Retrieved 24 June, 2020
- Four-major-types-of-earthquakes: inventionsky.com, Retrieved 27 August, 2020
- Earthquakes-prediction-9-methods-to-predict-earthquake-13915, earthquake: yourarticlelibrary.com, Retrieved 8 July, 2020
- Earthquakes-and-structural-damages, earthquakes-tectonics-hazard-and-risk-mitigation: intechopen.com, Retrieved 18 May, 2020
- Causes-and-Effects-of-Earthquake, stem: owlcation.com, Retrieved 9 May, 2020

2

Earthquake Measurement

The magnitude of earthquake can be measured through the use of seismometers, seismograph, Richter magnitude scale, accelerograph, modified Mercalli intensity scale, environmental seismic intensity scale, etc. The topics elaborated in this chapter will help in gaining a better perspective of these instruments used in measuring earthquake.

EARTHQUAKE INSTRUMENTATION

Earthquake detection instrumentation is widely used in many industries to secure the system, structure and components. In addition, early warning of incident seismic wave is critical to safety of human lives. On 12 September 2016, South Korea experienced the most powerful earthquake ever recorded in the country since measurements began in 1978. A 5.8-magnitude earthquake struck the historic city of Gyeongju and the people were subjected to a series of aftershocks affecting their daily lives. Buildings built inside the historic district were not strong enough to sustain seismic incident despite its low destructive power. Luckily, since the frequency of the wave was higher than 16 Hz, a catastrophic falling down of buildings was not experienced. This tells the relationship among magnitude, wave frequency and the destructive power. Earthquake detection instrumentation has been improved to detect the magnitude and frequency effectively. Historically, the first seismic instrument was invented in Han dynasty. Modern instruments are based on electronic sensors which are able to detect tri-axial acceleration.

In order to investigate both theory and application of earthquake instruments, the mathematical expression of the instrument is defined by the motion of the mass-spring-damper system. Then the differential equation resulting from the equilibrium of forces. In addition, the transfer functions of both force balanced and servo balanced and feedback circuit are analysed. Finally, the application of earthquake instrument in the nuclear power plant is introduced.

Mathematical Expression

Let us define a single seismic instrument having mass, spring and dashpot as shown in figure. A dashpot is a device that provides viscous friction or damping . The dashpot absorbs energy. This spring-mass-dashpot system can be expressed using Newton's second law as represented in equation:

$$(\text{D.E.})m\ddot{y} = -ky - b\dot{y} + u$$

$$(\text{T.F.})ms^2Y(s) = -kY(s) - bsY(s) + U(s)$$

$$G(s)\frac{Y(s)}{U(s)} = \frac{1}{ms^2 + bs + k}$$

The mathematical expression of the system above can be expressed as $(\text{D.E.})m\ddot{y} = -ky - b\dot{y} + u$. The transfer function of $(\text{D.E.})m\ddot{y} = -ky - b\dot{y} + u$ can be obtained as $(\text{T.F.})ms^2Y(s) = -kY(s) - bsY(s) + U(s)$ and $G(s)\frac{Y(s)}{U(s)} = \frac{1}{ms^2 + bs + k}$ by taking Laplace transformation.

The motion of the mass as a function of the ground displacement is given by a differential equation resulting from the equilibrium of forces as shown in equation:

$$Fs + Fr + Fg = 0$$

where $F_s = -k_x$ (k is the spring constant);

$F_r = -b_x$ (b is the friction coefficient);

$F_g = -m_u$ (m is the spring mass).

When the seismic instrument moves slowly, the acceleration and velocity become negligible, it records ground acceleration. If however, movement is fast enough, the acceleration of the mass dominates and the ground displacement is recorded.

Figure shows the seismic instrument under ground motion. Where the u_g is the ground displacement, x_r is displacement of the seismometer mass and x_o is the mass equilibrium position.

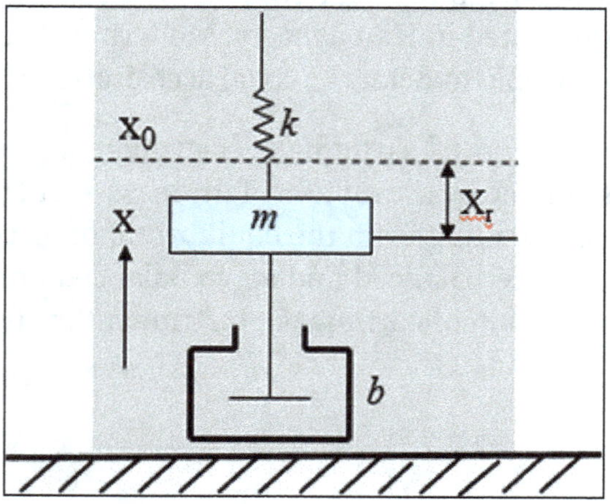

Spring-mass-dashpot system.

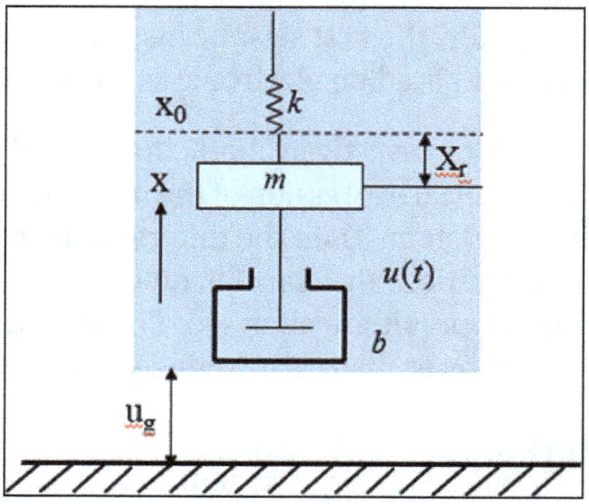

Concept of seismic instrument under ground motion.

In general, the frequency of the seismic wave tends to be increased if the magnitude of the wave is getting smaller. The ground motion has very low-frequency profile due to the movement of the earth.

Tri-axial Accelerometer

The tri-axial accelerometer has been widely used in industrial applications. An accelerometer is a sensor which can detect the earthquake-induced motion. The tri-axial accelerometer has a coil in the sensor that is suspended to it. When the accelerometer detects the earthquake-induced motion, the coil is moved and the voltage is induced proportionally to the variation of magnetic force. This voltage is converted to acceleration.

The tri-axial accelerometer detects incident seismic waves from longitudinal (L), transverse (T) and vertical (V) direction using three identical sensors as illustrated in figure. In general, each accelerometer has a feedback amplifier in consideration of the stability of the sensor output. Moreover, an added RC filter eliminates the high-frequency component from the output. The accelerometers are installed at free-field where the earthquake ground motion is detected by the surface motion.

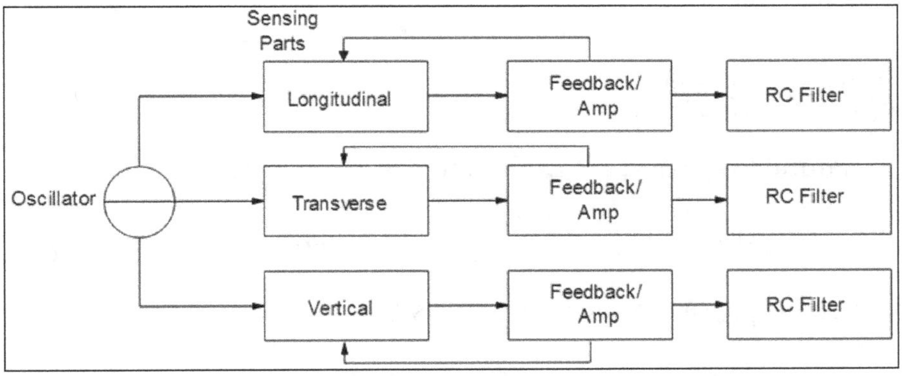

Concept of tri-axial accelerometer.

Types of Accelerometers

An accelerometer detects low-frequency component from the earthquake ground motion. Two measurement principles are used to detect a strong motion earthquake requiring high sensitivity: Open-loop type or closed-loop type with a feedback loop. In the plant industry, two sensor types have been used; force balanced acceleration using a pendulum, and servo acceleration with a feedback loop.

Force Balanced Acceleration using Pendulum

Here, the FBA-23 accelerometer is analysed to provide a more detailed operational theory of pendulous type servo sensors. As seen in figure, FBA-23 consists of three-axis FBAs (force balanced accelerators) for detecting incident seismic waves in (L), (T) and (V) directions. Each accelerometer is identical, however, the vertical one has an active oscillator circuit as an additional function.

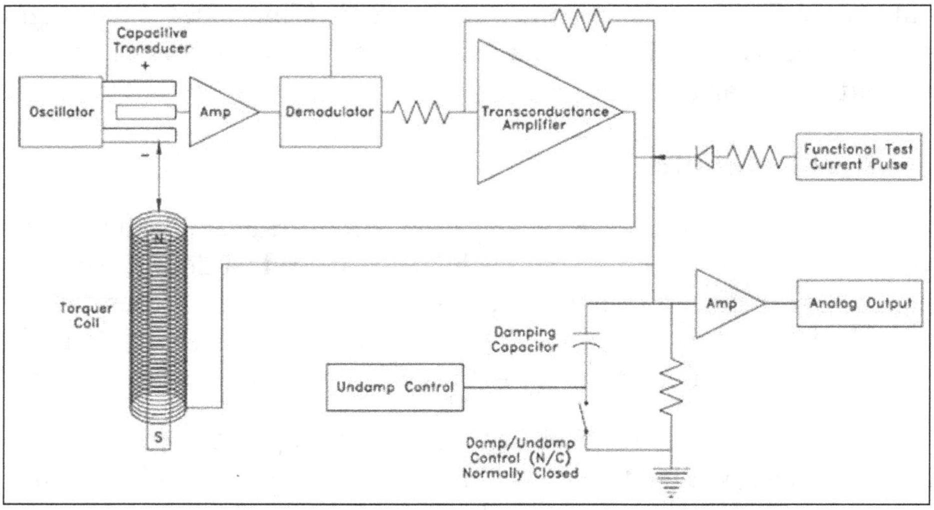

Simplified circuit of FBA-23.

The electrical circuit of FBA-23 has a non-inverting amplifying circuit for adjusting the polarity of the output signal. The transfer function of the accelerometer can be expressed by equation:

$$\frac{E_{\text{OUT}}}{E_{\text{DC}}} = \frac{\omega_n^2}{s^2 + 2\zeta\omega_n s + \omega_n^2}$$

where ω_n = natural frequency; s = Laplacian; ζ = damping factor.

The solution of $\frac{E_{\text{OUT}}}{E_{\text{DC}}} = \frac{\omega_n^2}{s^2 + 2\zeta\omega_n s + \omega_n^2}$ can be denoted by the second order equation having two poles ($p_{1,} p_2$) as shown in equations:

$$p_{1=} -\zeta\omega_n + j\omega_n\sqrt{1-\zeta}$$
$$p_{2=} -\zeta\omega_n - j\omega_n\sqrt{1-\zeta}$$

The circuit has sensitivities of 0.25, 0.5, 1, 2 and 4 g full scale. This corresponds to normal absolute sensitivity at DC.

The second order transfer function in $\frac{E_{\text{OUT}}}{E_{\text{DC}}} = \frac{\omega_n^2}{s^2 + 2\zeta\omega_n s + \omega_n^2}$ has natural frequencies of 50, 90 or 100 Hz. In addition, if we consider the damping factor of 0.707, then the two poles (p_1, p_2)

are located in the left half of the s-plane. The transfer function of this accelerometer can be rewritten as having two poles.

$$\frac{E_{\text{OUT}}}{E_{\text{DC}}} = \frac{\omega_n^2}{s^2 + 2\zeta\omega_n s + \omega_n^2} = \frac{\omega_n^2}{(s-P_1)(s-P_2)}$$

From $\frac{E_{OUT}}{E_{DC}} = \frac{\omega_n^2}{s^2 + 2\zeta\omega_n s + \omega_n^2} = \frac{\omega_n^2}{(s-P_1)(s-P_2)}$, the natural frequency and damping factor of this instrument can be calculated as,

$$\omega_n = 2 \bullet \pi \bullet 50 = 314.159$$

$$\zeta = 0.2550 \ (\text{for } 1.0 \text{ g})$$

The poles are then calculated as, P_1 = -222.1 + j222.1 rad/s and P_2 = -222.1 - j222.1 rad/s. Therefore, the resultant transfer function of this system becomes,

$$TF = \frac{98696}{s^2 + 160.22s + 98696}$$

Figure shows the Matlab Simulink results when the step input is applied. The corresponding plot described in figure shows overshoots and it needs relatively reduced settling time. In this simulation, the poles have 50 Hz of natural frequency and 1 g of absolute gain.

Matlab Simulink diagram (without RC filter).

Since this transfer function may be affected by the zeros, therefore by adding low pass RC filter to the post amplifier, the stability and accuracy of this circuit would be improved. In this case, the RC filter adds one more pole as summarized in table. This proves that the calculated poles are placed in the left half of the s-plane as well.

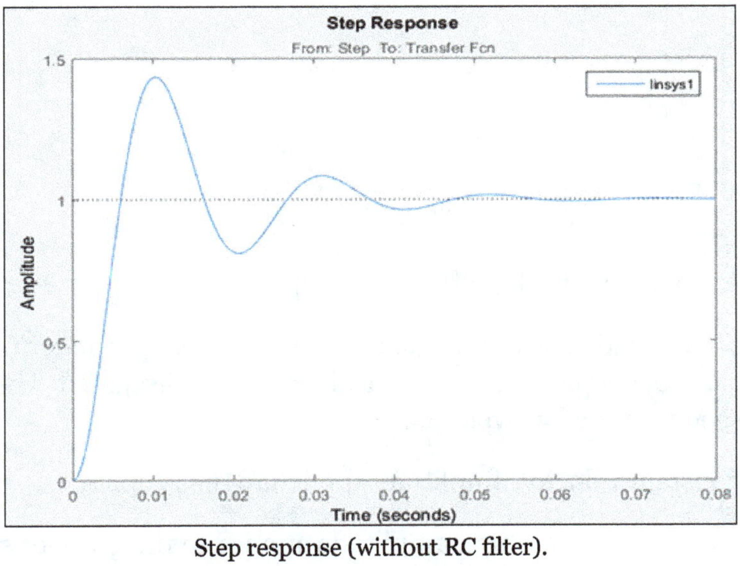

Step response (without RC filter).

Table: Pole placement of FBA-23 in 50 Hz and 1 g.

	P_1	P_2	P_3
Real (rad/s)	−222.1	−222.1	−1000
Imaginary (rad/s)	222.1	−222.1	0

The full transfer function of the electrical circuit in 50 Hz has three poles as shown in equation:

$$TF_{RC} = \frac{98696}{s^3 + 1160.225 s^2 + 258921 s + 98696000}$$

Figure shows the Matlab Simulink results when a step input is applied. Figure also shows almost the same plot disregarding the existence of the RC filter.

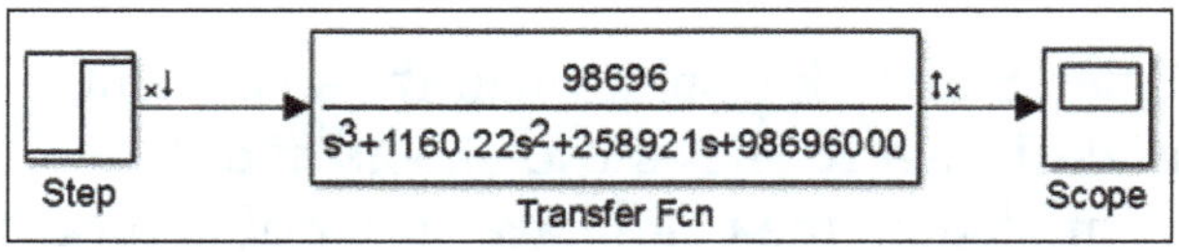

Matlab Simulink diagram (with RC filter).

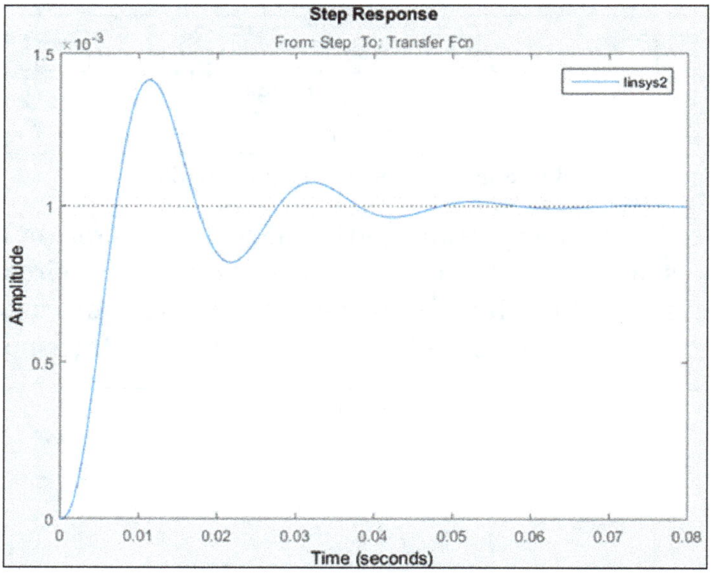

Step response (with RC filter).

Servo Acceleration with Feedback Loop

Here, the servo acceleration with a feedback loop type accelerometer (AC-23 by Geosig), which consists of geophone and electrical circuits, is discussed. The geophone response can be seen from two viewpoints:

- With a constant velocity, amplitude applied to the geophone.
- With a constant acceleration, amplitude applied to the geophone.

Figure represents the response curve of Geophone at 4.5 Hz.

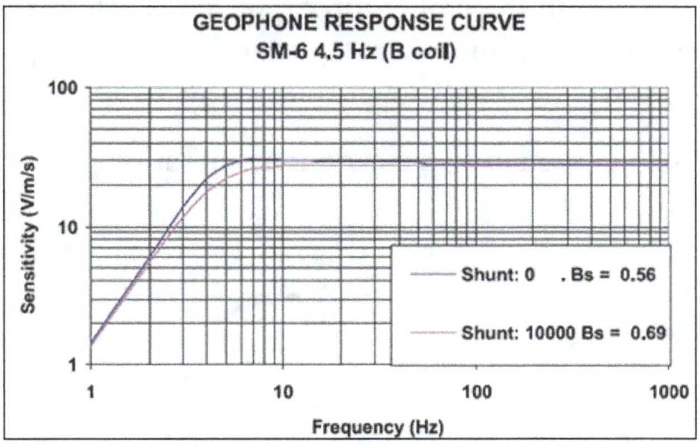

Geophone response curve (4.5Hz).

The relationship between these two points of view is a rotation when plotted with log-log axis scales. In figure, the principle of Geophone is drawn. When the motion on the x-axis is incident then Geophone generates the open-loop signal which has peak response of 4.5 Hz.

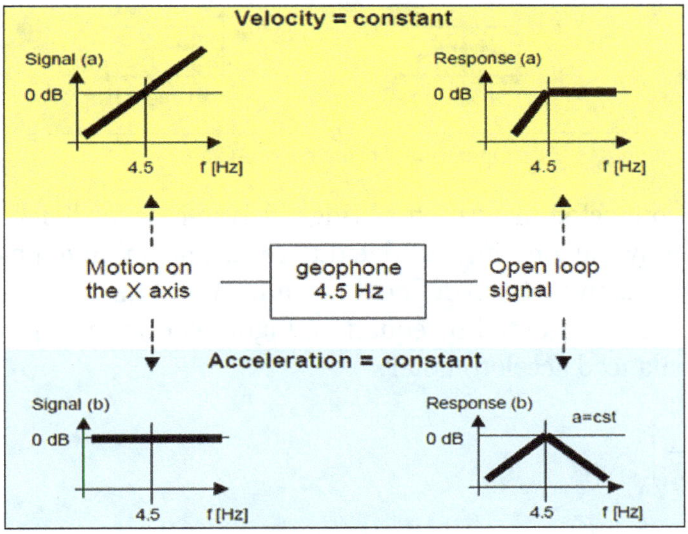

Principle of Geophone.

The geophone is a long coil travel version and the extra coil travel offers an advantage for higher tilt requirements where larger amplitude signals may be encountered. A range of natural frequencies is available from 4.5 Hz to 14 Hz, providing a choice of the correct geophone for a wide variety of applications. The geophone is suitable for detecting an extensive range of ground motion.

The AC-23 provides over-damping the geophones with a feedback amplifier in a bridge circuit. The principle of over-damping geophones is done by applying a voltage on the geophone, which has opposite polarity from the voltage, which is induced by the moving geophone coil. Since the voltage induced by the geophone coil is proportional to the velocity, the externally applied voltage is also proportional to it. This results in a current in the coil (~force), which is also proportional to the velocity and therefore is a "damping" current, or additional damping. Increasing this damping further will lead to a resulting output that is proportional to acceleration.

The function of the sensor, seen from a constant acceleration point of view is illustrated in figure.

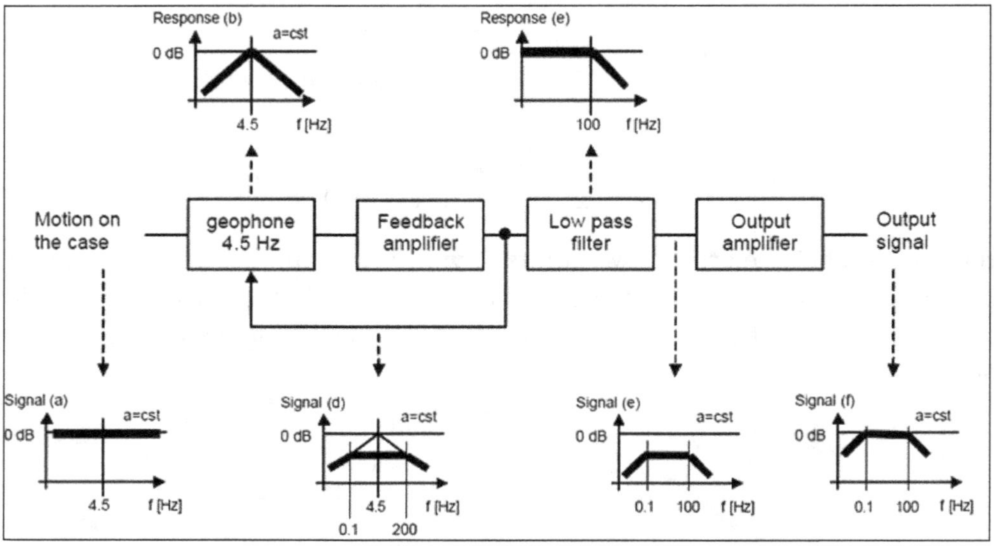

Operational principle of AC-23.

The geophone is connected in a resistor bridge, driven by a feedback amplifier and an inverter, which apply the amplified bridge differential signal in opposite polarity. The amplifier has a fixed gain that will define the bandwidth of the accelerometer. The gain G and output V_o are represented by equation. Figure shows the equivalent circuit for the linear servo-balanced accelerometer.

$$G = 1 + \left(\frac{Z_2}{Z_1}\right)$$

$$V_0 = G \bullet V_i = \left(1 + \frac{Z_2}{Z_1}\right) \bullet V_i$$

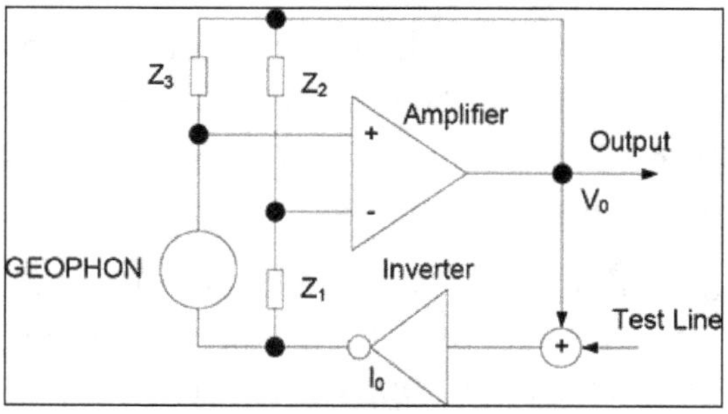

Equivalent circuit for AC-23 (linear servo balanced accelerometer).

In figure, the test-line shifts the voltage to one side of the bridge, which produces a current flow in the geophone, resulting in a displacement step of the seismic mass. The movement of the mass generates a voltage across the Geophone, which is detected by the differential amplifier and induces an output signal. The effect of the test signal on the bridge is cancelled by the differential input of the amplifier.

In this part, the Geophone transfer function transfer is analysed in order to define how the sensor generates open-loop transfer function as,

$$TF = \frac{1.564s}{0.1251s^2 + 1.514s + 1}$$

The transfer function in $TF = \frac{1.564s}{0.1251s^2 + 1.514s + 1}$ has two poles and one zero. In this work, the MATLAB script of Geophone is used to illustrate the response curve under step input condition as shown in figure.

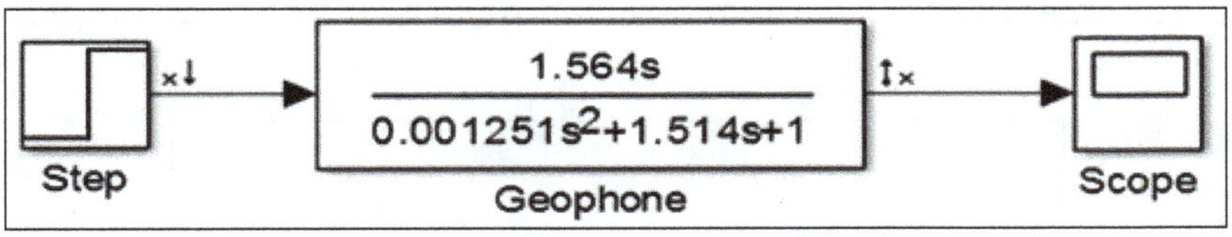

Open-loop transfer function of Geophone.

The transfer function plotted as the frequency range vs. relative response to the acceleration is shown in figure.

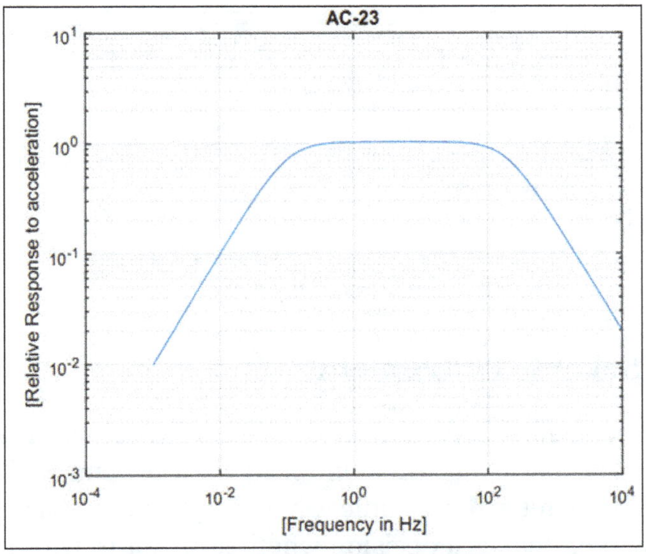

Transfer function plotted by the frequency vs. relative response to the acceleration.

If a unit step function is applied to the transfer function, the response as in figure is obtained.

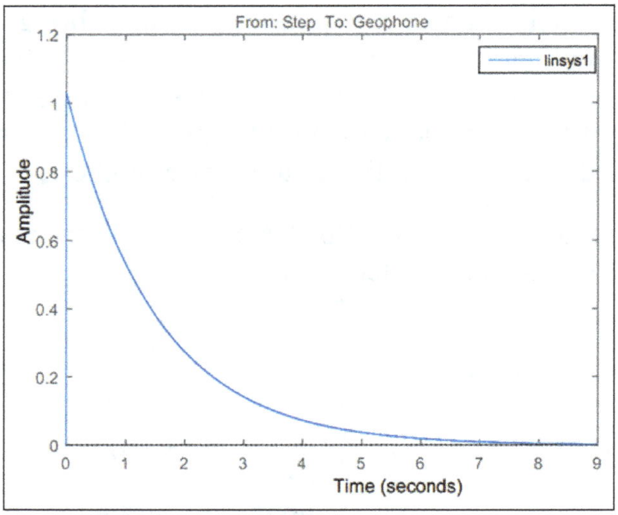

Step response.

In this case, the magnitude and the phase can be drawn by the Bode plot as seen in figure.

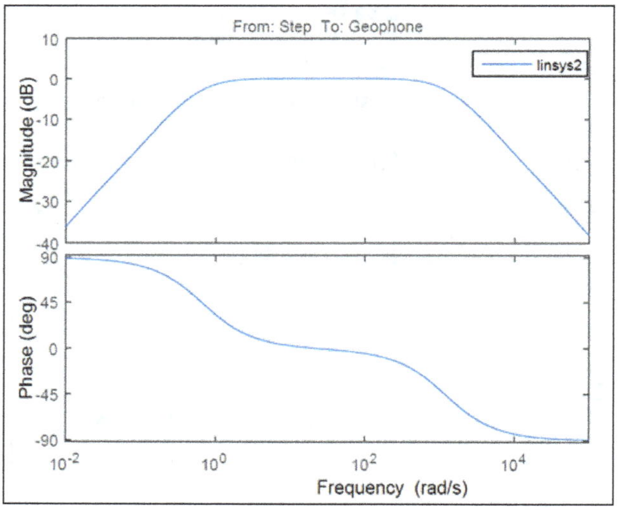

Bode plot.

Application of Seismic Instrument

Here, two application examples for the NPP (nuclear power plant) are discussed with their designs. They are SMS (seismic monitoring system) and ASTS (automatic seismic trip system). SMS detects the seismic wave and provides alarms to the NPP operators. The operator then makes a decision whether or not to trip the nuclear reactor. ASTS has an automatic trip function, if the peak acceleration of the incident wave exceeds, the pre-determined level trigger the reactor trip automatically. In the protection view-point, SMS provides alarms for the manual trip while ASTS provides automatic protection.

Seismic Monitoring System

The SMS consists of tri-axial accelerometers and a seismic monitoring cabinet which is located in an electrical equipment room in unit one (only one unit). The cabinet contains recorders, a playback and analysis unit, an annunciator, a seismic switch, test units and an uninterruptible power supply.

The signal flow from the accelerometer to the seismic monitoring system is shown in figure. From the accelerometer, the voltage signal is transmitted to the seismic monitoring cabinet and is converted by electronics into a signal proportional to the acceleration. The acceleration signal can activate recorders and/or local and MCR annunciators/alarms according to its magnitude.

When the acceleration of earthquake exceeds the seismic trigger or set-points for OBE (Operating Basis Earthquake or Event) or SSE (Safe Shutdown Earthquake), the alarm is provided on the local annunciator panel of the seismic monitoring cabinet and that of large panel display system (LDPS) in main control room or alternatively in the remote shutdown room. That event also activates the time-history recorder(s) in the seismic monitoring cabinet.

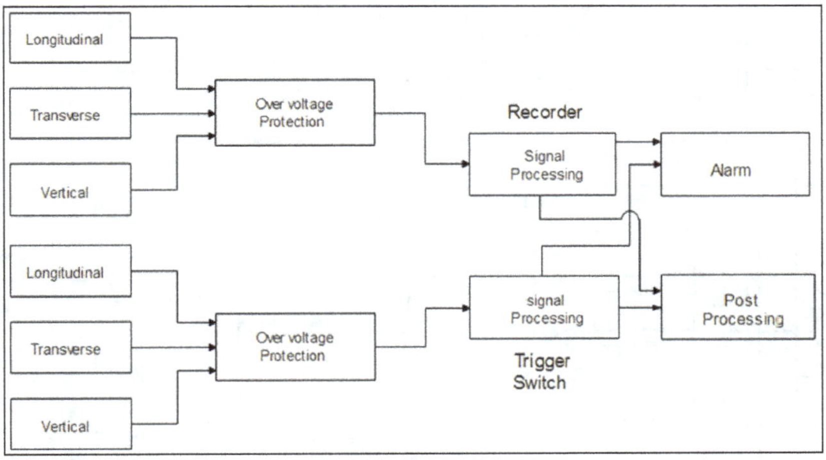

SMS block diagram.

Automatic Seismic Trip System

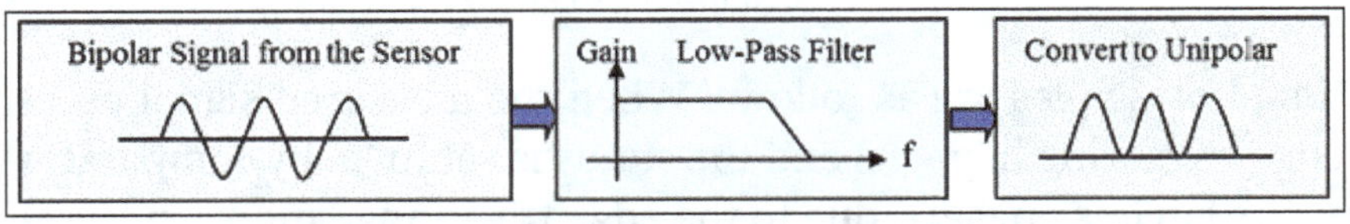

Conceptual diagram of sensor module for signal processing.

ASTS continuously monitor PGA (peak ground acceleration) from the seismic wave and automatically generates the trip signal. As shown in figure, the seismic signal detected by the accelerometer is filtered, rectified and converted to the current. The sensor module is to cut off the frequency ranges over 10 Hz in order to pick out the strong motion of the earthquake. Since the seismic signal has a positive and negative signal, a rectifier circuit is added. The rectifier converts the bipolar signal to the unipolar.

The overall scheme of ASTS is shown in figure. The sensor output is interfaced with digital input card at the ASTS cabinet. Even the decision logic is set as two out of four but the system is composed of two independent channels. They are N_1 and N_2. For isolation between the channels, the digital input/output card is applied. The dotted line on the ASTS cabinet is the scope of trip logic channel N_1 and N_2. The ASTS cabinet is implemented by digitalized system such as PLC (programmable logic controller), FPGA (field programmable gate array), or DCS (distribute control system). The bi-stable and decision logic are configured with software inside each channel.

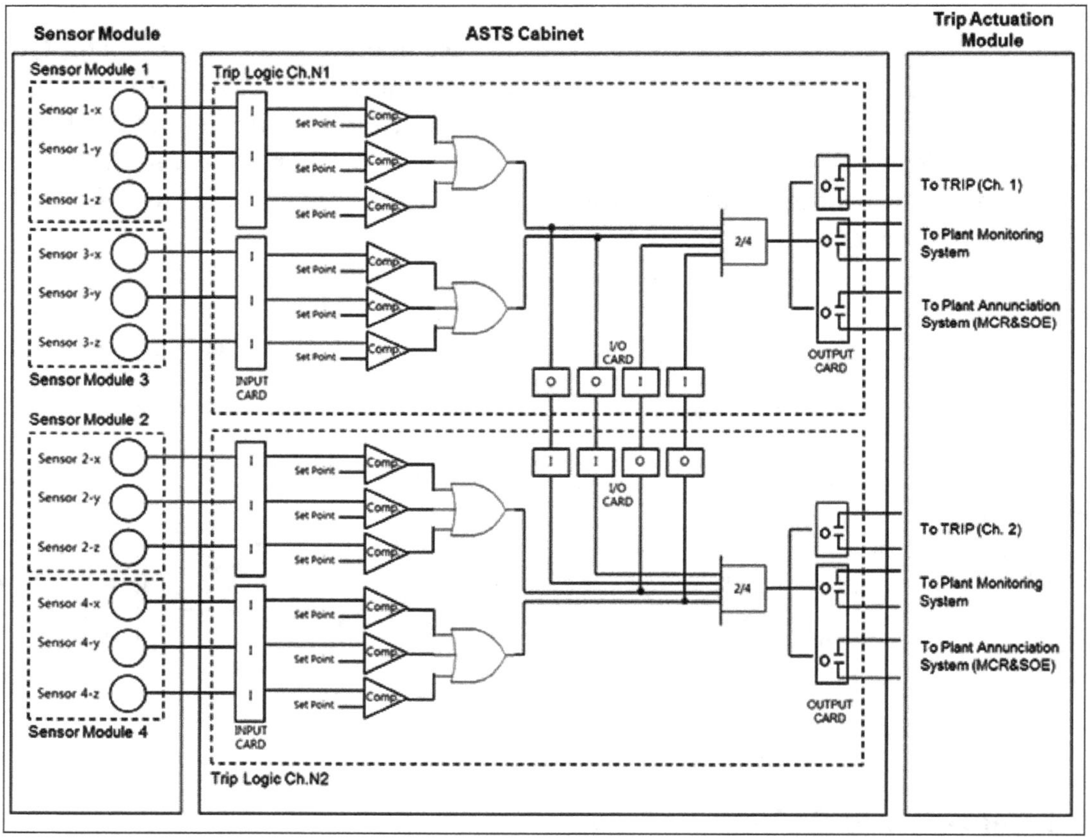

Digitalized ASTS.

The operational principles are as follows. When the measured signal exceeds the set-point, the edge triggering happens and the status is set to '0' by comparator actuation. In this case, the latch is engaged due to the edge triggering, and keep the status for 10 s. The purpose of 10-^{s} latch is to synchronize the individual channel for the coincidence logic. If the signals are not properly synchronized, the trip initiation signal may not be actuated during a strong earthquake. ASTS consists of four diverse channel applications. The bi-stable signal from each channel feeds into two-out of-four coincidence logic for generating trip initiation signal.

SEISMOMETERS

A seismometer is a device that is sensitive to vibrations. It works on the principle of a pendulum: A heavy, inert mass with a certain resistance to movement (i.e. inertia) due to its weight is suspended from a frame by a spring that allows movement. The energy from any seismic activity excites this "proof mass" as it is called by geophysicists, making it vibrate.

Simple Pendulums and Inverted Pendulums

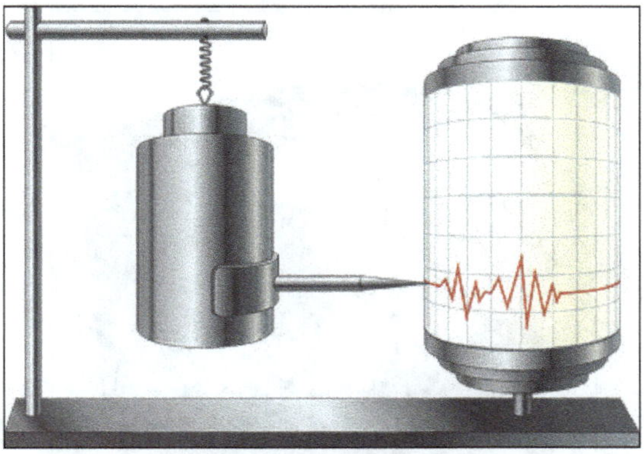

Principle of operation of a seismometer.

The first seismometers developed were based on a simple pendulum in which the moving mass is suspended vertically from a frame.

To increase sensitivity, this type of device was then mounted upside down, which is why it is known as an inverted pendulum. In this case, the pendulum's centre of mass is above the pivot point.

Unlike the first design, this assembly is naturally unstable and the least disturbance will make that mass leave its point of equilibrium and move left or right as gravity attracts it downwards.

The inherent instability of an inverted pendulum nonetheless is what makes this device react to the slightest movement, however small.

Plotting a Seismogram

The seismometer provides data as a seismogram, which is a recording of the magnitude of ground motion over time while regularly measuring the offset between the position of the mass and the frame to which it is attached, in relation to an equilibrium position i.e. when the device is at rest in the absence of any seismic activity.

Historically, the first seismograms were produced by basic means: a metal needle attached to a mobile mass left a track of its movements on paper covered in soot. Later, the needle was replaced by an ink pen that recorded movements on a rotating roll of paper. Nowadays, in the digital age, seismometers provide digital signals recorded by computers. The seismometers flown to the Moon by the Apollo missions were among the first of this kind, because even in the early 1970s, most seismic stations on Earth were still fitted with seismographs.

The role of modern sensors is to follow the motion of the mass, and they can take three types of measurement: changes in the position of the mass relative to its "zero" equilibrium position (displacement), speed of the mass (i.e. motion over time), and finally a variation in speed over time (i.e. acceleration or deceleration).

Recording of a seismic signal.

Some types of sensor are more appropriate than others depending on what type of measurement you wish to make. Inductive sensors are geared to measuring speed, while capacitive sensors are ideal for measuring the position of the mass.

Number of Axes

As ground motion occurs in three-dimensional space (along the vertical axis or two horizontal axes), displacements need to be recorded using three separate pendulums. This is the only way that seismic activity can be comprehensively documented.

Sophisticated seismometers usually measure all three axes, one for each direction in space. Simple one-axis seismometers usually only measure vertical motion; horizontal surface motion is ignored since it is not measured.

The SEIS instrument designed to fly aboard. An interesting detail is that the three axes of the seismometer are not aligned with the horizontal and vertical, which would perhaps be "logical".

Although each of SEIS's three axes is positioned at a 90° angle to the others, the whole assembly is inclined by approximately 30.5° with respect to the horizontal axis. There are many, sometimes complex, reasons for this but to simplify matters, let us just say that such a configuration minimizes the effect of noise.

The sensitivity of a seismometer depends on the relationship between the seismic signal that you wish to record and the various disturbances that interfere with this signal. This "noise", which can go as far as preventing measurements, is similar to the snow effect you can see on a television screen.

The 30.5° angle of Very BroadBand (VBB) pendulums corresponds to an angle where the vertical component is measured with the best sensitivity relative to the instrument's own noise. Tipping it towards the horizontal axis would reduce the amplitude of vertical acceleration, while tipping it away would increase the instrument's self-noise, with increasing oscillation leading to a loss in long-period sensitivity.

Ideal Set-up Conditions

One of the issues to face when using a seismometer is how to ensure that it can optimally monitor a vibration even when the vibration lasts a long time, from minutes to hours.

Seismometers are by definition extremely sensitive, recording all that is going on around them whether seismic or not. However, the slightest drift of any kind can prevent a seismometer from continuing to monitor and plot a vibration, especially if the latter is a long-period event (the recording of short-period events being easier to handle).

Geophysicists attach great importance to the way seismometers are set up. However sensitive an instrument is, it will only give good results if it is set up correctly.

Seismometers are generally placed on very hard surfaces such as granite or concrete slabs. The surface must be as flat and as horizontal as possible.

Ideally, the place chosen must be quiet. It is not a good idea to set a seismometer up next to a road or near an underground railway station. Temperature is also very important. It must be as stable as possible, as any variation may affect the seismometer mechanics, especially the force generated by the spring on the mass.

On Earth, seismometers are preferably set up in the pits or shafts of former mines, hundreds of metres below the surface. These are good locations as they provide ideal operating conditions.

The sources of interfering noise, which propagate very easily near the surface (such as the continual hum of human activities, the constant roar of the oceans, and atmospheric turbulence), are reduced as far as possible. As for temperature, it is incredibly and naturally stable. The quietest site in Europe is in the Black Forest, in the workings of a former mine. The temperature there only varies by a few thousandths of a degree per year.

Once in an ideal location, a seismometer can feel and monitor any seismic wave, whether very short or, on the contrary, lasting a matter of minutes or even hours.

Seismic Wave Sources

A seismometer is designed to record seismic waves. On Earth, these waves are mainly caused by earthquakes that regularly and sometimes forcefully shake certain parts of the world, causing major damage depending on the amount of energy they carry.

Apart from major earthquakes that make newspaper headlines, the Earth's surface is constantly subject to very small movements that we do not detect but which are revealed on seismograms.

Our planet can actually act like a bell, and when certain events occur, can even resonate in rhythm. The Earth then produces its own music, a telluric melody imperceptible to our ears but able to be picked up by the most sensitive seismometers.

Whether the seismic waves have been released during major seismic activity that shakes certain regions of the world or whether they are just due to the imperceptible oscillations of our planet, geophysicists use every opportunity to broaden our knowledge of the Earth's interior.

A seismometer is rather like a doctor's stethoscope. By determining the way that seismic waves propagate within our planet, depending on how they are reflected or refracted by the materials making up the planet's interior structure, whether rocks or metal, it becomes possible to construct an image.

Passive and Active Sounding

When it is necessary to study a particular sub-surface area when prospecting for oil, for instance, you cannot always count on an earthquake at the right time. Seismic activity does not occur on demand.

There are two solutions in this case: either you listen for long periods of time in the hope that seismic activity will generate seismic waves as input for the seismometers, or you create your own vibrations. The first is known as passive sounding, the second as active sounding.

SEISMOGRAM

A seismogram is a graph output by a seismograph. It is a record of the ground motion at a measuring station as a function of time. Seismograms typically record motions in three cartesian axes (x, y, and z), with the z axis perpendicular to the Earth's surface and the x- and y- axes parallel to the surface. The energy measured in a seismogram may result from an earthquake or from some other source, such as an explosion. Seismograms can record lots of things, and record many little waves, called microseisms. These tiny microseisms can be caused by heavy traffic near the seismograph, waves hitting a beach, the wind, and any number of other ordinary things that cause some shaking of the seismograph.

A seismogram being recorded by a seismograph.

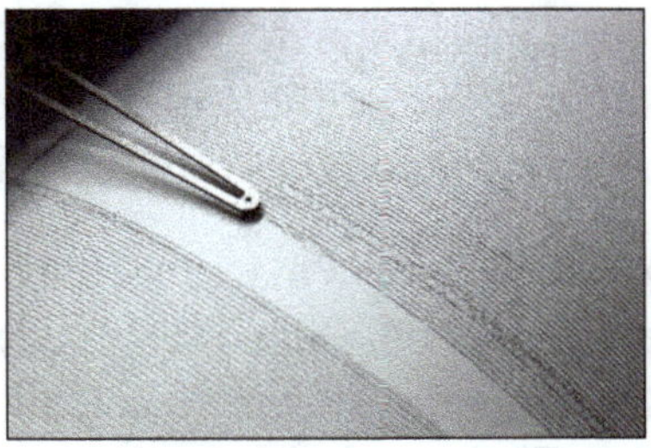

A detail of the seismogram.

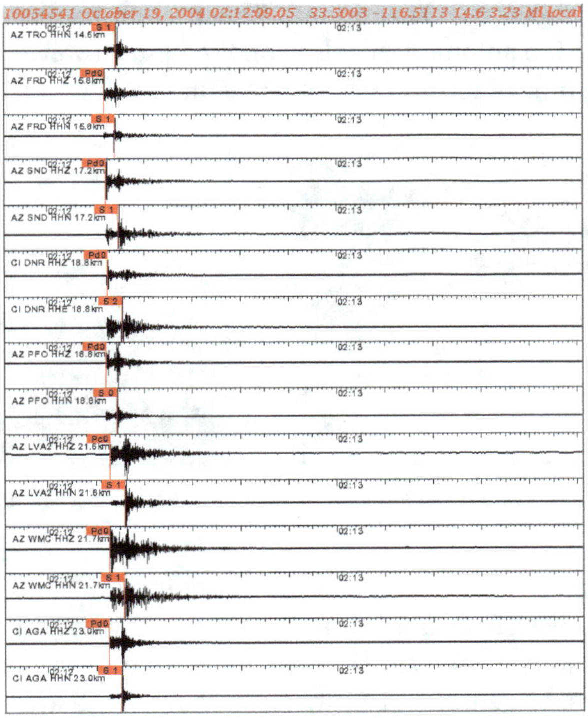

A set of seismograms for an earthquake from the USGS.

Historically, seismograms were recorded on paper attached to rotating drums. Some used pens on ordinary paper, while others used light beams to expose photosensitive paper. Today, practically all seismograms are recorded digitally to make analysis by computer easier. Some drum seismometers are still found, especially when used for public display. Seismograms are essential for finding the location and magnitude of earthquakes.

Recording

Prior to the availability of digital processing of seismic data in the late 1970s, the records were done in a few different forms on different types of media.

A Helicorder drum is a device used to record data into photographic paper or in the form of paper and ink. A piece of paper is wrapped around a rotating drum of the helicorder which receives the seismic signal from a seismometer. For each predefined interval of data, the helicorder will plot the seismic data in one line before moving to the next line at the next interval. The paper must be changed after the helicorder writes on the last line of the paper. In the model that use ink, regular maintenance of the pen must be done for accurate recording.

A Develocorder is a machine that record multi-channels seismic data into a 16 mm film. The machine was developed by Teledyne Geotech during the mid 1960s. It can automatically plot seismograms from 18 seismic signal sources and 3 time signals on a continuous reel of film. The signals from seismometers are processed by 15.5 Hz recording galvanometers which record the seismograms to a reel of 200 feet (61 m) of film at the speeds between 3 and 20 centimetres (1.2 and 7.9 in) per minute. The machine has self-contained circulating chemicals that are used to automatically develop the film. However, the machine takes at least 10 minutes from time of recording to the time that the film can be viewed.

A container that stores a Develocorder film reel.

A Develocorder film reel.

After the digital processing has been used, the archives of the seismograms were recorded in magnetic tapes. The data from the magnetic tapes can then be read back to reconstruct the original waveforms. Unfortunately, due to the deterioration of older magnetic tape medias, large number of waveforms from the archives in the early digital recording days are not recoverable. Today, many other forms are used to digitally record the seismograms into digital medias.

Reading

Seismograms are read from left to right. Time marks show when the earthquake occurred. Time is shown by half-hour (thirty-minute) units. Each rotation of the seismograph drum is thirty minutes. Therefore, on seismograms, each line measures thirty minutes. This is a more efficient way to read a seismogram. Secondly, there are the minute-marks. A minute mark looks like a hyphen "-" between each minute. Minute marks count minutes on seismograms. From left to right, each mark stands for a minute.

Each seismic wave looks different. The P-wave is the first wave that is bigger than the other waves (the microseisms). Because P waves are the fastest seismic waves, they will usually be the first ones that the seismograph records. The next set of seismic waves on the seismogram will be the S-waves. These are usually bigger than the P waves, and have higher frequency. Look for a dramatic change in frequency for a different type of wave.

SEISMIC INTENSITY SCALES

Seismic intensity scales categorize the intensity or severity of ground shaking (quaking) at a given location, such as resulting from an earthquake. They are distinguished from seismic magnitude scales, which measure the magnitude or overall strength of an earthquake, which may, or perhaps not, cause perceptible shaking.

Intensity scales are based on the observed effects of the shaking, such as the degree to which people or animals were alarmed, and the extent and severity of damage to different kinds of structures or natural features. The maximal intensity observed, and the extent of the area where shaking was felt, can be used to estimate the location and magnitude of the source earthquake; this is especially useful for historical earthquakes where there is no instrumental record.

Ground Shaking

Ground shaking can be caused in various ways (volcanic tremors, avalanches, large explosions, etc.), but shaking intense enough to cause damage is usually due to rupturing of the earth's crust known as earthquakes. The intensity of shaking depends on several factors:

- The "size" or strength of the source event, such as measured by various seismic magnitude scales.
- The type of seismic wave generated, and its orientation.
- The depth of the event.
- The distance from the source event.
- Site response due to local geology.

Site response is especially important as certain conditions, such as unconsolidated sediments in a basin, can amplify ground motions as much as ten times.

Where an earthquake is not recorded on seismographs an isoseismal map showing the intensities felt at different areas can be used to estimate the location and magnitude of the quake. Such maps are also useful for estimating the shaking intensity, and thereby the likely level of damage, to be expected from a future earthquake of similar magnitude. In Japan this kind of information is used when an earthquake occurs to anticipate the severity of damage to be expected in different areas.

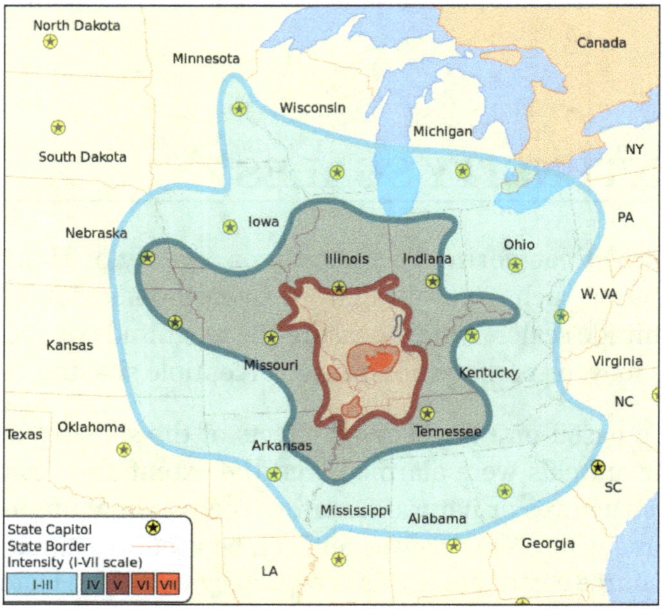

Isoseismal map for the Illinois earthquake, showing the extent of different levels of shaking. The irregularity of areas is due to ground conditions and the underlying geology.

The intensity of local ground-shaking depends on several factors besides the magnitude of the earthquake, one of the most important being soil conditions. For instance, thick layers of soft soil (such as fill) can amplify seismic waves, often at a considerable distance from the source, while sedimentary basins will often resonate, increasing the duration of shaking. This is why, in the 1989 Loma Prieta earthquake, the Marina district of San Francisco was one of the most damaged areas, though it was nearly 100 km from the epicenter. Geological structures were also significant, such as where seismic waves passing under the south end of San Francisco Bay reflected off the base of the Earth's crust towards San Francisco and Oakland. A similar effect channeled seismic waves between the other major faults in the area.

MODIFIED MERCALLI INTENSITY SCALE

The Modified Mercalli intensity scale (MM or MMI is a seismic intensity scale used for measuring the intensity of shaking produced by an earthquake. It measures the effects of an earthquake at a given location, distinguished from the earthquake's inherent force or strength as measured by seismic magnitude scales (such as the "M_w" magnitude usually reported for an earthquake). While shaking is driven by the seismic energy released by an earthquake, earthquakes differ in how much of their energy is radiated as seismic waves. Deeper earthquakes also have less interaction with the surface, and their energy is spread out across a larger area. Shaking intensity is localized, generally diminishing with distance from the earthquake's epicenter, but can be amplified in sedimentary basins and certain kinds of unconsolidated soils.

Intensity scales empirically categorize the intensity of shaking based on the effects reported by untrained observers, and are adapted for the effects that might be observed in a particular region. In not requiring instrumental measurements, they are useful for estimating the magnitude and location of historical (pre-instrumental) earthquakes: the greatest intensities generally correspond to the epicentral area, and their degree and extent (possibly augmented with knowledge of local geological conditions) can be compared with other local earthquakes to estimate the magnitude.

The lower degrees of the Modified Mercalli Intensity scale generally deal with the manner in which the earthquake is felt by people. The higher numbers of the scale are based on observed structural damage.

This table gives Modified Mercalli scale intensities that are typically observed at locations near the epicenter of the earthquake.

I. Not felt	Not felt except by very few under especially favorable conditions.
II. Weak	Felt only by a few people at rest, especially on upper floors of buildings.
III. Weak	Felt quite noticeably by people indoors, especially on upper floors of buildings. Many people do not recognize it as an earthquake. Standing motor cars may rock slightly. Vibrations similar to the passing of a truck. Duration estimated.
IV. Light	Felt indoors by many, outdoors by few during the day. At night, some awakened. Dishes, windows, doors disturbed; walls make cracking sound. Sensation like heavy truck striking building. Standing motor cars rocked noticeably.
V. Moderate	Felt by nearly everyone; many awakened. Some dishes, windows broken. Unstable objects overturned. Pendulum clocks may stop.
VI. Strong	Felt by all, many frightened. Some heavy furniture moved; a few instances of fallen plaster. Damage slight.
VII. Very strong	Damage negligible in buildings of good design and construction; slight to moderate in well-built ordinary structures; considerable damage in poorly built or badly designed structures; some chimneys broken.
VIII. Severe	Damage slight in specially designed structures; considerable damage in ordinary substantial buildings with partial collapse. Damage great in poorly built structures. Fall of chimneys, factory stacks, columns, monuments, walls. Heavy furniture overturned.

IX. Violent	Damage considerable in specially designed structures; well-designed frame structures thrown out of plumb. Damage great in substantial buildings, with partial collapse. Buildings shifted off foundations. Liquefaction.
X. Extreme	Some well-built wooden structures destroyed; most masonry and frame structures destroyed with foundations. Rails bent.
XI. Extreme	Few, if any, (masonry) structures remain standing. Bridges destroyed. Broad fissures in ground. Underground pipe lines completely out of service. Earth slumps and land slips in soft ground. Rails bent greatly.
XII. Extreme	Damage total. Waves seen on ground surfaces. Lines of sight and level distorted. Objects thrown upward into the air.

Correlation with Magnitude

Magnitude	Magnitude/intensity comparison
1.0–3.0	I
3.0–3.9	II–III
4.0–4.9	IV–V
5.0–5.9	VI–VII
6.0–6.9	VII–IX
7.0 and higher	VIII or higher

The correlation between magnitude and intensity is far from total, depending upon several factors including the depth of the hypocenter, terrain, distance from the epicenter. For example, a 4.5 magnitude quake in Salta, Argentina, in 2011, that was 164 km deep had a maximum intensity of I, while a 2.2 magnitude event in Barrow in Furness, England, in 1865, about 1 km deep had a maximum intensity of VIII.

Estimating Site Intensity and its use in Seismic Hazard Assessment

Dozens of so-called intensity prediction equations have been published to estimate the macroseismic intensity at a location given the magnitude, source-to-site distance and, perhaps, other parameters (e.g. local site conditions). These are similar to ground motion prediction equations for the estimation of instrumental strong-motion parameters such as peak ground acceleration. Such equations can be used to estimate the seismic hazard in terms of macroseismic intensity, which has the advantage of being more closely related to seismic risk than instrumental strong-motion parameters.

Correlation with Physical Quantities

The Mercalli scale is not defined in terms of more rigorous, objectively quantifiable measurements such as shake amplitude, shake frequency, peak velocity, or peak acceleration. Human-perceived shaking and building damages are best correlated with peak acceleration for lower-intensity events, and with peak velocity for higher-intensity events.

Comparison to the Moment Magnitude Scale

The effects of any one earthquake can vary greatly from place to place, so there may be many Mercalli intensity values measured for the same earthquake. These values can be best displayed using a contoured map of equal intensity, known as an isoseismal map. However, each earthquake has only one magnitude.

ENVIRONMENTAL SEISMIC INTENSITY SCALE

The Environmental Seismic Intensity scale (ESI 2007) is a seismic scale used for measuring the intensity of an earthquake on the basis of the effects of the earthquake on the natural environment (Earthquake Environmental Effects).

The international effort to develop a new scale of macroseismic intensity that would focus exclusively on environmental effects of earthquakes began in the early 1990s and was sponsored by the International Union for Quaternary Research (INQUA). After the final draft of the scale was approved by INQUA at its XVII Congress in Cairns, Australia, in 2007, the scale became officially known as ESI 2007.

Like many other intensity scales, ESI 2007 uses the basic structure of twelve degrees of seismic intensity and is designed for application during field surveys immediately after the seismic event. However, the definitions of intensity degrees in ESI 2007 are based on the observation of distribution and size of environmental effects produced by an earthquake. This approach makes ESI 2007 a unique diagnostic tool for the assessment of seismic intensity levels X to XII in sparsely populated and uninhibited areas where earthquake effects on people and built environment may not be easily observed. For intensity level IX or lower, the ESI 2007 scale is intended to be used as a supplement to other intensity scales:

- The Definition of intensity degrees on the basis of Earthquake Environmental Effects.
- The Guidelines, which aim at better clarifying i) the background of the scale and the scientific concepts that support the introduction of such a new macroseismic scale; ii) the procedure to use the scale alone or integrated with damage-based, traditional scales; iii) how the scale is organized; iv) the descriptions of diagnostic features required for intensity assessment, and the meaning of idioms, colors, and fonts.

Since 2007, the scale has been applied to modern, historical and paleoearthquakes in the frame of the INQUA TERPRO Commission on Paleoseismology and Active tectonics. It is now available in ten different languages.

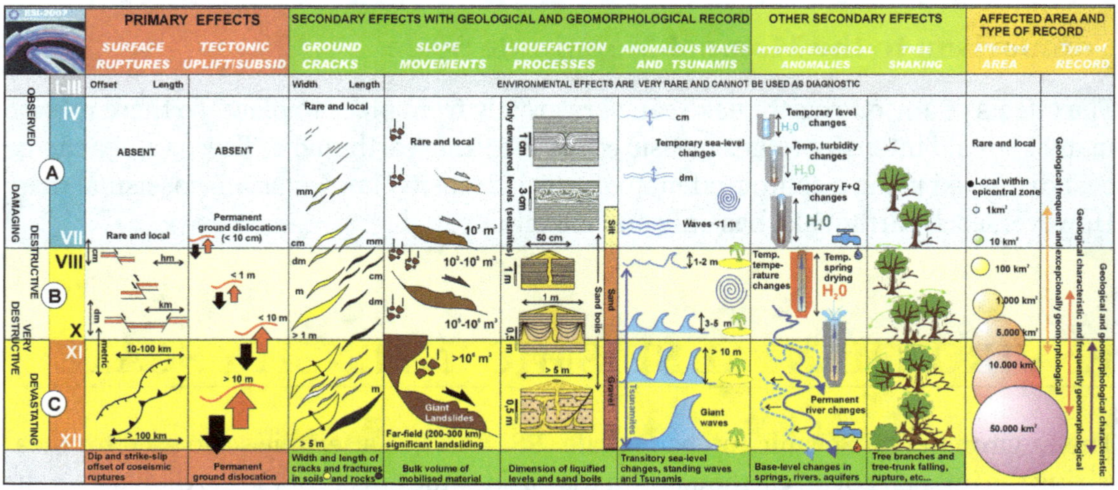

SEISMIC MAGNITUDE SCALES

Seismic magnitude scales are used to describe the overall strength or "size" of an earthquake. These are distinguished from seismic intensity scales that categorize the intensity or severity of ground shaking (quaking) caused by an earthquake at a given location. Magnitudes are usually determined from measurements of an earthquake's seismic waves as recorded on a seismogram. Magnitude scales vary on what aspect of the seismic waves are measured and how they are measured. Different magnitude scales are necessary because of differences in earthquakes, the information available, and the purposes for which the magnitudes are used.

Earthquake Magnitude and Ground-shaking Intensity

The Earth's crust is stressed by tectonic forces. When this stress becomes great enough to rupture the crust, or to overcome the friction that prevents one block of crust from slipping past another, energy is released, some of it in the form of various kinds of seismic waves that cause ground-shaking, or quaking.

Magnitude is an estimate of the relative "size" or strength of an earthquake, and thus its potential for causing ground-shaking. It is "approximately related to the released seismic energy."

Intensity refers to the strength or force of shaking at a given location, and can be related to the peak ground velocity. With an isoseismal map of the observed intensities an earthquake's magnitude can be estimated from both the maximum intensity observed (usually but not always near the epicenter), and from the extent of the area where the earthquake was felt.

The intensity of local ground-shaking depends on several factors besides the magnitude of the earthquake, one of the most important being soil conditions. For instance, thick layers of soft soil (such as fill) can amplify seismic waves, often at a considerable distance from the source, while sedimentary basins will often resonate, increasing the duration of shaking. This is why, in the 1989 Loma Prieta earthquake, the Marina district of San Francisco was one of the most damaged areas, though it was nearly 100 km from the epicenter. Geological structures were also significant, such as where seismic waves passing under the south end of San Francisco Bay reflected off the base of the Earth's crust towards San Francisco and Oakland. A similar effect channeled seismic waves between the other major faults in the area.

Magnitude Scales

An earthquake radiates energy in the form of different kinds of seismic waves, whose characteristics reflect the nature of both the rupture and the earth's crust the waves travel through. Determination of an earthquake's magnitude generally involves identifying specific kinds of these waves on a seismogram, and then measuring one or more characteristics of a wave, such as its timing, orientation, amplitude, frequency, or duration. Additional adjustments are made for distance, kind of crust, and the characteristics of the seismograph that recorded the seismogram.

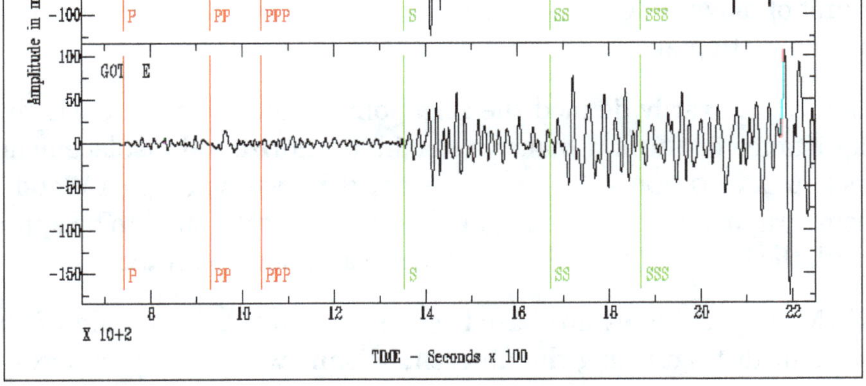

Typical seismogram: The compressive P-waves (following the red lines) – essentially sound passing through rock – are the fastest seismic waves, and arrive first, typically in about 10 seconds for an earthquake around 50 km away. The sideways-shaking S-waves (following the green lines) arrive some seconds later, traveling a little over half the speed of the P-waves; the delay is a direct indication of the distance to the quake. S-waves may take an hour to reach a point 1000 km away. Both of these are body-waves, that pass directly through the earth's crust. Following the S-waves are various kinds of surface-waves – Love waves and Rayleigh waves – that travel only at the earth's surface. Surface waves are smaller for deep earthquakes, which have less interaction with the surface. For shallow earthquakes – less than roughly 60 km deep – the surface waves are stronger, and may last several minutes; these carry most of the energy of the quake, and cause the most severe damage.

The various magnitude scales represent different ways of deriving magnitude from such information as is available. All magnitude scales retain the logarithmic scale as devised by Charles Richter, and are adjusted so the mid-range approximately correlates with the original "Richter" scale.

Most magnitude scales are based on measurements of only part of an earthquake's seismic wave-train, and therefore are incomplete. This results in systematic underestimation of magnitude in certain cases, a condition called saturation.

Since 2005 the International Association of Seismology and Physics of the Earth's Interior (IASPEI) has standardized the measurement procedures and equations for the principal magnitude scales, M_L, M_s, mb, mB and mb_{Lg}.

"Richter" Magnitude Scale

The first scale for measuring earthquake magnitudes, developed in 1935 by Charles F. Richter and popularly known as the "Richter" scale, is actually the Local magnitude scale, label ML or M_L. Richter established two features now common to all magnitude scales. First, the scale is logarithmic, so that each unit represents a ten-fold increase in the amplitude of the seismic waves. As the energy of a wave is $10^{1.5}$} times its amplitude, each unit of magnitude represents a nearly 32-fold increase in the seismic energy (strength) of an earthquake.

Second, Richter arbitrarily defined the zero point of the scale to be where an earthquake at a distance of 100 km makes a maximum horizontal displacement of 0.001 millimeters (1 μm, or 0.00004 in.) on a seismogram recorded with a Wood-Anderson torsion seismograph. Subsequent magnitude scales are calibrated to be approximately in accord with the original "Richter" (local) scale around magnitude 6.

All "Local" (ML) magnitudes are based on the maximum amplitude of the ground shaking, without distinguishing the different seismic waves. They underestimate the strength:

- Of distant earthquakes (over ~600 km) because of attenuation of the S-waves.

- Of deep earthquakes because the surface waves are smaller.
- Of strong earthquakes (over M ~7) because they do not take into account the duration of shaking.

The original "Richter" scale, developed in the geological context of Southern California and Nevada, was later found to be inaccurate for earthquakes in the central and eastern parts of the continent (everywhere east of the Rocky Mountains) because of differences in the continental crust. All these problems prompted the development of other scales.

Most seismological authorities, such as the United States Geological Survey, report earthquake magnitudes above 4.0 as moment magnitude, which the press describes as "Richter magnitude".

Other "Local" Magnitude Scales

Richter's original "local" scale has been adapted for other localities. These may be labelled "ML", or with a lowercase "l", either Ml, or M_l. Whether the values are comparable depends on whether the local conditions have been adequately determined and the formula suitably adjusted.

Japanese Meteorological Agency Magnitude Scale

In Japan, for shallow (depth < 60 km) earthquakes within 600 km, the Japanese Meteorological Agency calculates a magnitude labeled MJMA, M_{JMA}, or M_J. (These should not be confused with moment magnitudes JMA calculates, which are labeled M_w(JMA) or $M^{(JMA)}$, nor with the Shindo intensity scale.) JMA magnitudes are based (as typical with local scales) on the maximum amplitude of the ground motion; they agree "rather well" with the seismic moment magnitude M_w in the range of 4.5 to 7.5, but underestimate larger magnitudes.

Body-wave Magnitude Scales

Body-waves consist of P-waves that are the first to arrive, or S-waves, or reflections of either. Body-waves travel through rock directly.

mB Scale

The original "body-wave magnitude" – mB or m_B (uppercase "B") – was developed by Gutenberg and Richter to overcome the distance and magnitude limitations of the M_L scale inherent in the use of surface waves. mB is based on the P- and S-waves, measured over a longer period, and does not saturate until around M 8. However, it is not sensitive to events smaller than about M 5.5. Use of mB as originally defined has been largely abandoned, now replaced by the standardized mB_{BB} scale.

mb Scale

The m_b or m_b scale (lowercase "m" and "b") is similar to mB, but uses only P-waves measured in the first few seconds on a specific model of short-period seismograph. It was introduced in the 1960s with the establishment of the World-Wide Standardized Seismograph Network (WWSSN); the short period improves detection of smaller events, and better discriminates between tectonic earthquakes and underground nuclear explosions.

Measurement of mb has changed several times. As originally defined by Gutenberg m_b was based on the maximum amplitude of waves in the first 10 seconds or more. However, the length of the period influences the magnitude obtained. Early USGS/NEIC practice was to measure mb on the first second (just the first few P-waves), but since 1978 they measure the first twenty seconds. The modern practice is to measure short-period mb scale at less than three seconds, while the broadband mB_{BB} scale is measured at periods of up to 30 seconds.

mb_{Lg} Scale

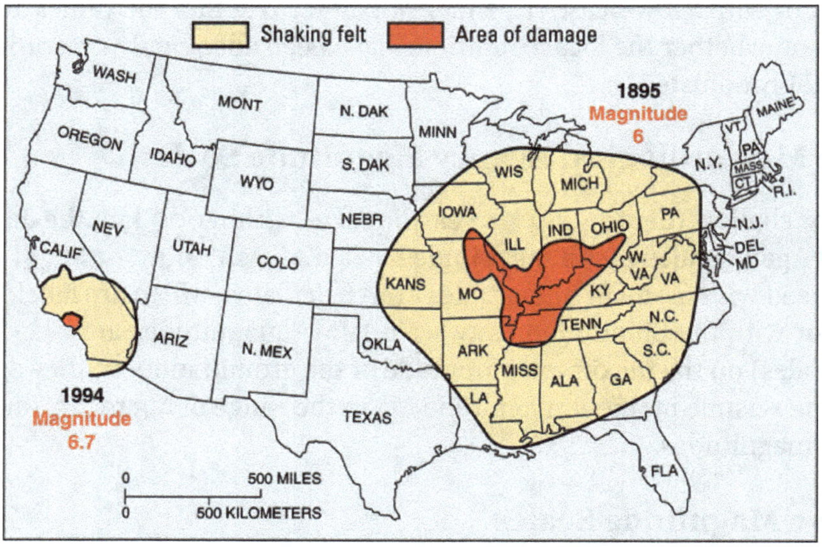

Differences in the crust underlying North America east of the Rocky Mountains makes that area more sensitive to earthquakes.

The regional mb_{Lg} scale – also denoted mb_Lg, mbLg, MLg (USGS), Mn, and m_N – was developed by Nuttli for a problem the original M_L scale could not handle: all of North America east of the Rocky Mountains. The M_L scale was developed in southern California, which lies on blocks of oceanic crust, typically basalt or sedimentary rock, which have been accreted to the continent. East of the Rockies the continent is a craton, a thick and largely stable mass of continental crust that is largely granite, a harder rock with different seismic characteristics. In this area the M_L scale gives anomalous results for earthquakes which by other measures seemed equivalent to quakes in California.

Nuttli resolved this by measuring the amplitude of short-period (~1 sec.) Lg waves, a complex form of the Love wave which, although a surface wave, he found provided a result more closely related the m_b scale than the M_s scale. Lg waves attenuate quickly along any oceanic path, but propagate well through the granitic continental crust, and Mb_{Lg} is often used in areas of stable continental crust; it is especially useful for detecting underground nuclear explosions.

Surface-wave Magnitude Scales

Surface waves propagate along the Earth's surface, and are principally either Rayleigh waves or Love waves. For shallow earthquakes the surface waves carry most of the energy of the earthquake, and are the most destructive. Deeper earthquakes, having less interaction with the surface, produce weaker surface waves.

The surface-wave magnitude scale, variously denoted as Ms, M_S, and M_s, is based on a procedure developed by Beno Gutenberg in 1942 for measuring shallow earthquakes stronger or more distant than Richter's original scale could handle. Notably, it measured the amplitude of surface waves (which generally produce the largest amplitudes) for a period of "about 20 seconds". The M_s scale approximately agrees with M_L at ~6, then diverges by as much as half a magnitude. A revision by Nuttli , sometimes labeled M_{Sn}, measures only waves of the first second.

A modification – the "Moscow-Prague formula" – was proposed in 1962, and recommended by the IASPEI in 1967; this is the basis of the standardized M_{s20} scale (Ms_20, M_s(20)). A "broad-band" variant (Ms_BB, M_s(BB)) measures the largest velocity amplitude in the Rayleigh-wave train for periods up to 60 seconds. The M_{S7} scale used in China is a variant of M_s calibrated for use with the Chinese-made "type 763" long-period seismograph. The MLH scale used in some parts of Russia is actually a surface wave magnitude.

Moment Magnitude and Energy Magnitude Scales

Other magnitude scales are based on aspects of seismic waves that only indirectly and incompletely reflect the force of an earthquake, involve other factors, and are generally limited in some respect of magnitude, focal depth, or distance. The moment magnitude scale – Mw or M_w is based on an earthquake's seismic moment, M_0, a measure of how much work an earthquake does in sliding one patch of rock past another patch of rock. Seismic moment is measured in Newton-meters (N • m or Nm) in the SI system of measurement, or dyne-centimeters (dyn-cm) in the older CGS system. In the simplest case the moment can be calculated knowing only the amount of slip, the area of the surface ruptured or slipped, and a factor for the resistance or friction encountered. These factors can be estimated for an existing fault to determine the magnitude of past earthquakes, or what might be anticipated for the future.

An earthquake's seismic moment can be estimated in various ways, which are the bases of the M_{wb}, M_{wr}, M_{wc}, M_{ww}, M_{wp}, M_i, and M_{wpd} scales, all subtypes of the generic M_w scale.

Seismic moment is considered the most objective measure of an earthquake's "size" in regard of total energy. However, it is based on a simple model of rupture, and on certain simplifying assumptions; it incorrectly assumes that the proportion of energy radiated as seismic waves is the same for all earthquakes.

Much of an earthquake's total energy as measured by M_w is dissipated as friction (resulting in heating of the crust). An earthquake's potential to cause strong ground shaking depends on the comparatively small fraction of energy radiated as seismic waves, and is better measured on the energy magnitude scale, M_e. The proportion of total energy radiated as seismic waves varies greatly depending on focal mechanism and tectonic environment; M_e and M_w for very similar earthquakes can differ by as much as 1.4 units.

Despite the usefulness of the M_e scale, it is not generally used due to difficulties in estimating the radiated seismic energy.

Energy Class (K-class) Scale

K is a measure of earthquake magnitude in the energy class or K-class system, developed in 1955 by Soviet seismologists in the remote Garm (Tadjikistan) region of Central Asia; in revised form it is still used for local and regional quakes in many states formerly aligned with the Soviet Union (including Cuba). Based on seismic energy ($K = \log E_S$, in Joules), difficulty in implementing it using the technology of the time led to revisions in 1958 and 1960. Adaptation to local conditions has led to various regional K scales, such as K_F and K_S.

K values are logarithmic, similar to Richter-style magnitudes, but have a different scaling and zero point. K values in the range of 12 to 15 correspond approximately to M 4.5 to 6. M(K), $M_{(K)}$, or possibly M_K indicates a magnitude M calculated from an energy class K.

Tsunami Magnitude Scales

Earthquakes that generate tsunamis generally rupture relatively slowly, delivering more energy at longer periods (lower frequencies) than generally used for measuring magnitudes. Any skew in the spectral distribution can result in larger, or smaller, tsunamis than expected for a nominal magnitude. The tsunami magnitude scale, M_t, is based on a correlation by Katsuyuki Abe of earthquake seismic moment (M_0) with the amplitude of tsunami waves as measured by tidal gauges. Originally intended for estimating the magnitude of historic earthquakes where seismic data is lacking but tidal data exist, the correlation can be reversed to predict tidal height from earthquake magnitude. Under low-noise conditions, tsunami waves as little as 5 cm can be predicted, corresponding to an earthquake of M ~6.5.

Another scale of particular importance for tsunami warnings is the mantle magnitude scale, M_m. This is based on Rayleigh waves that penetrate into the Earth's mantle, and can be determined quickly, and without complete knowledge of other parameters such as the earthquake's depth.

Duration and Coda Magnitude Scales

M_d designates various scales that estimate magnitude from the duration or length of some part of the seismic wave-train. This is especially useful for measuring local or regional earthquakes, both powerful earthquakes that might drive the seismometer off-scale (a problem with the analog instruments formerly used) and preventing measurement of the maximum wave amplitude, and weak earthquakes, whose maximum amplitude is not accurately measured. Even for distant earthquakes, measuring the duration of the shaking (as well as the amplitude) provides a better measure of the earthquake's total energy. Measurement of duration is incorporated in some modern scales, such as M_{wpd} and mB_c.

M_c scales usually measure the duration or amplitude of a part of the seismic wave, the coda. For short distances (less than ~100 km) these can provide a quick estimate of magnitude before the quake's exact location is known.

Macroseismic Magnitude Scales

Magnitude scales generally are based on instrumental measurement of some aspect of the seismic wave as recorded on a seismogram. Where such records do not exist, magnitudes can be estimated from reports of the macroseismic events such as described by intensity scales.

One approach for doing this (developed by Beno Gutenberg and Charles Richter in 1942) relates the maximum intensity observed (presumably this is over the epicenter), denoted I_0 (capital I, subscripted zero), to the magnitude. It has been recommended that magnitudes calculated on this basis be labeled $M_w(I_0)$, but are sometimes labeled with a more generic M_{ms}.

Another approach is to make an isoseismal map showing the area over which a given level of intensity was felt. The size of the "felt area" can also be related to the magnitude. While the recommended label for magnitudes derived in this way is $M_0(An)$, the more commonly seen label is M_{fa}. A variant, M_{La}, adapted to California and Hawaii, derives the Local magnitude (M_L) from the size of the area affected by a given intensity. M_I (upper-case letter "I", distinguished from the lower-case letter in M_i) has been used for moment magnitudes estimated from isoseismal intensities calculated per Johnston 1996.

Peak ground velocity (PGV) and Peak ground acceleration (PGA) are measures of the force that causes destructive ground shaking. In Japan, a network of strong-motion accelerometers provides PGA data that permits site-specific correlation with different magnitude earthquakes. This correlation can be inverted to estimate the ground shaking at that site due to an earthquake of a given magnitude at a given distance. From this a map showing areas of likely damage can be prepared within minutes of an actual earthquake.

Other Magnitude Scales

Many earthquake magnitude scales have been developed or proposed, with some never gaining broad acceptance and remaining only as obscure references in historical catalogs of earthquakes. Other scales have been used without a definite name, often referred to as "the method of Smith " (or similar language), with the authors often revising their method. On top of this, seismological networks vary on how they measure seismograms. When the details of how a magnitude has been determined are unknown, catalogs will specify the scale as unknown (variously Unk, Ukn, or UK). In such cases, the magnitude is considered generic and approximate.

An M_h ("magnitude determined by hand") label has been used where the magnitude is too small or the data too poor (typically from analog equipment) to determine a Local magnitude, or multiple shocks or cultural noise complicates the records. The Southern California Seismic Network uses this "magnitude" where the data fail the quality criteria.

RICHTER MAGNITUDE SCALE

Richter scale or Richter's magnitude scale – for measuring the strength ("size") of earthquakes refers to the original "magnitude scale" developed by Charles F. Richter and presented in his landmark 1935 paper. This was later revised and renamed the Local magnitude scale, denoted as "ML" or "M_L". Because of various shortcomings of the ML scale most seismological authorities now use other scales, such as the moment magnitude scale (M_w), to report earthquake magnitudes, but much of the news media still refers to these as "Richter" magnitudes. All magnitude scales retain the logarithmic character of the original, and are scaled to have roughly comparable numeric values.

The Richter scale was defined in 1935 for particular circumstances and instruments; the particular circumstances refer to it being defined for Southern California and "implicitly incorporates the attenuative properties of Southern California crust and mantle." The particular instrument used would become saturated by strong earthquakes and unable to record high values. The scale was replaced in the 1970s by the moment magnitude scale (MMS, symbol M_w); for earthquakes adequately measured by the Richter scale, numerical values are approximately the same. Although values measured for earthquakes now are M_w (MMS), they are frequently reported by the press as Richter values, even for earthquakes of magnitude over 8, when the Richter scale becomes meaningless.

The Richter and MMS scales measure the energy released by an earthquake; another scale, the Mercalli intensity scale, classifies earthquakes by their effects, from detectable by instruments but not noticeable, to catastrophic. The energy and effects are not necessarily strongly correlated; a shallow earthquake in a populated area with soil of certain types can be far more intense in effects than a much more energetic deep earthquake in an isolated area.

Several scales have historically been described as the "Richter scale", especially the local magnitude M_L and the surface wave M_s scale. In addition, the body wave magnitude, m_b, and the moment magnitude, M_w, abbreviated MMS, have been widely used for decades. A couple of new techniques to measure magnitude are in the development stage by seismologists.

All magnitude scales have been designed to give numerically similar results. This goal has been achieved well for M_L, M_s, and M_w. The m_b scale gives somewhat different values than the other scales. The reason for so many different ways to measure the same thing is that at different distances, for different hypocentral depths, and for different earthquake sizes, the amplitudes of different types of elastic waves must be measured.

M_L is the scale used for the majority of earthquakes reported (tens of thousands) by local and regional seismological observatories. For large earthquakes worldwide, the moment magnitude scale (MMS) is most common, although M_s is also reported frequently.

The seismic moment, M_o, is proportional to the area of the rupture times the average slip that took place in the earthquake, thus it measures the physical size of the event. M_w is derived from it empirically as a quantity without units, just a number designed to conform to the M_s scale. A spectral analysis is required to obtain M_o, whereas the other magnitudes are derived from a simple measurement of the amplitude of a specifically defined wave.

All scales, except M_w, saturate for large earthquakes, meaning they are based on the amplitudes of waves which have a wavelength shorter than the rupture length of the earthquakes. These short waves (high frequency waves) are too short a yardstick to measure the extent of the event. The resulting effective upper limit of measurement for M_L is about 7 and about 8.5 for M_s.

New techniques to avoid the saturation problem and to measure magnitudes rapidly for very large earthquakes are being developed. One of these is based on the long period P-wave; the other is based on a recently discovered channel wave.

The energy release of an earthquake, which closely correlates to its destructive power, scales with the $\frac{3}{2}$ power of the shaking amplitude. Thus, a difference in magnitude of 1.0 is equivalent to a factor of 31.6 ($=(10^{1.0})^{(3/2)}$) in the energy released; a difference in magnitude of 2.0 is equivalent to a factor of 1000 ($=(10^{2.0})^{(3/2)}$) in the energy released. The elastic energy radiated is best derived from an integration of the radiated spectrum, but an estimate can be based on m_b because most energy is carried by the high frequency waves.

Richter Magnitudes

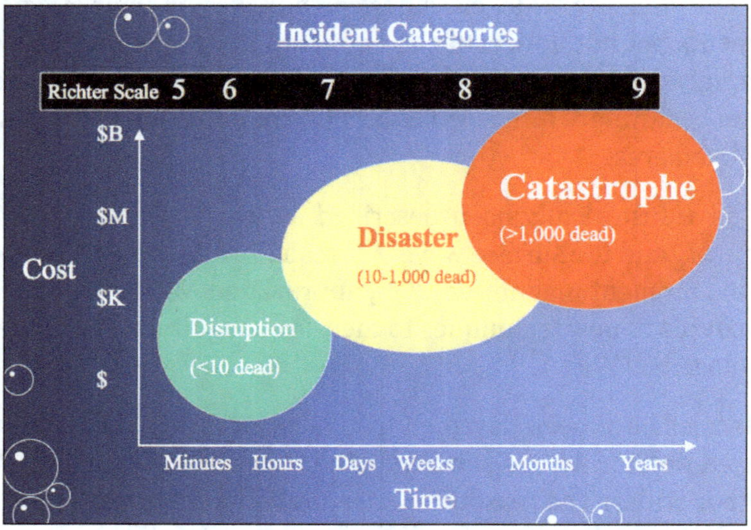

The Richter magnitude of an earthquake is determined from the logarithm of the amplitude of waves recorded by seismographs (adjustments are included to compensate for the variation in the distance between the various seismographs and the epicenter of the earthquake). The original formula is:

$$M_L = \log_{10} A - \log_{10} A_0(\delta) = \log_{10}[A / A_0(\delta)],$$

where A is the maximum excursion of the Wood–Anderson seismograph, the empirical function A_0 depends only on the epicentral distance of the station, δ. In practice, readings from all observing stations are averaged after adjustment with station-specific corrections to obtain the M_L value.

Because of the logarithmic basis of the scale, each whole number increase in magnitude represents a tenfold increase in measured amplitude; in terms of energy, each whole number increase corresponds to an increase of about 31.6 times the amount of energy released, and each increase of 0.2 corresponds to approximately a doubling of the energy released.

Events with magnitudes greater than 4.5 are strong enough to be recorded by a seismograph anywhere in the world, so long as its sensors are not located in the earthquake's shadow.

The following describes the typical effects of earthquakes of various magnitudes near the epicenter. The values are typical only. They should be taken with extreme caution, since intensity and thus ground effects depend not only on the magnitude, but also on the distance to the epicenter, the depth of the earthquake's focus beneath the epicenter, the location of the epicenter and geological conditions (certain terrains can amplify seismic signals).

Magnitude	Description	Mercalli intensity	Average earthquake effects	Average frequency of occurrence globally (estimated)
1.0–1.9	Micro	I	Microearthquakes, not felt, or felt rarely. Recorded by seismographs.	Continual/several million per year
2.0–2.9	Minor	I to II	Felt slightly by some people. No damage to buildings.	Over one million per year
3.0–3.9		III to IV	Often felt by people, but very rarely causes damage. Shaking of indoor objects can be noticeable.	Over 100,000 per year
4.0–4.9	Light	IV to VI	Noticeable shaking of indoor objects and rattling noises. Felt by most people in the affected area. Slightly felt outside. Generally causes none to minimal damage. Moderate to significant damage very unlikely. Some objects may fall off shelves or be knocked over.	10,000 to 15,000 per year
5.0–5.9	Moderate	VI to VII	Can cause damage of varying severity to poorly constructed buildings. At most, none to slight damage to all other buildings. Felt by everyone.	1,000 to 1,500 per year
6.0–6.9	Strong	VIII to X	Damage to a moderate number of well-built structures in populated areas. Earthquake-resistant structures survive with slight to moderate damage. Poorly designed structures receive moderate to severe damage. Felt in wider areas; up to hundreds of miles/kilometers from the epicenter. Strong to violent shaking in epicentral area.	100 to 150 per year
7.0–7.9	Major	X or greater	Causes damage to most buildings, some to partially or completely collapse or receive severe damage. Well-designed structures are likely to receive damage. Felt across great distances with major damage mostly limited to 250 km from epicenter.	10 to 20 per year
8.0–8.9	Great		Major damage to buildings, structures likely to be destroyed. Will cause moderate to heavy damage to sturdy or earthquake-resistant buildings. Damaging in large areas. Felt in extremely large regions.	One per year
9.0 and greater			At or near total destruction – severe damage or collapse to all buildings. Heavy damage and shaking extends to distant locations. Permanent changes in ground topography.	One per 10 to 50 years

The intensity and death toll depend on several factors (earthquake depth, epicenter location, and population density, to name a few) and can vary widely.

Minor earthquakes occur every day and hour. On the other hand, great earthquakes occur once a year, on average. The largest recorded earthquake was the Great Chilean earthquake of May 22, 1960, which had a magnitude of 9.5 on the moment magnitude scale. The larger the magnitude, the less frequently the earthquake happens.

Seismologist Susan Hough has suggested that 10 may represent a very approximate upper limit, as the effect if the largest known continuous belt of faults ruptured together (along the Pacific coast of the Americas).

Magnitude Empirical Formulae

These formulae for Richter magnitude M_L are alternatives to using Richter correlation tables based on Richter standard seismic event ($M_L = 0,\ A = 0.001mm,\ D = 100km$). Below, Δ is the epicentral distance (in kilometers unless otherwise specified).

The Lillie empirical formula:

$$M_L = \log_{10} A - 2.48 + 2.76 \log_{10} \Delta,$$

Where A is the amplitude (maximum ground displacement) of the P-wave, in micrometers, measured at 0.8 Hz.

For distances D less than 200 km,

$$M_L = \log_{10} A + 1.6 \log_{10} D - 0.15,$$

and for distances between 200 km and 600 km,

$$M_L = \log_{10} A + 3.0 \log_{10} D - 3.38,$$

where A is seismograph signal amplitude in mm and D is in km.

The Bisztricsany empirical formula for epicentral distances between 4° to 160°:

$$M_L = 2.92 + 2.25 \log_{10}(\tau) - 0.001\Delta^{\circ},$$

Where τ is the duration of the surface wave in seconds, and Δ is in degrees. M_L is mainly between 5 and 8.

The Tsumura empirical formula:

$$M_L = -2.53 + 2.85 \log_{10}(F - P) + 0.0014\Delta^{\circ}$$

Where $F-P$ is the total duration of oscillation in seconds. M_L is mainly between 3 and 5.

The Tsuboi, University of Tokyo, empirical formula:

$$M_L = \log_{10} A + 1.73 \log_{10} \Delta - 0.83$$

Where A is the amplitude in micrometers.

PEAK GROUND ACCELERATION

Peak ground acceleration (PGA) is equal to the maximum ground acceleration that occurred during earthquake shaking at a location. PGA is equal to the amplitude of the largest absolute acceleration recorded on an accelerogram at a site during a particular earthquake. Earthquake shaking generally occurs in all three directions. Therefore, PGA is often split into the horizontal and vertical components. Horizontal PGAs are generally larger than those in the vertical direction but this is not always true, especially close to large earthquakes. PGA is an important parameter (also known as an intensity measure) for earthquake engineering, The design basis earthquake ground motion (DBEGM) is often defined in terms of PGA.

Unlike the Richter and moment magnitude scales, it is not a measure of the total energy (magnitude, or size) of an earthquake, but rather of how hard the earth shakes at a given geographic point. The Mercalli intensity scale uses personal reports and observations to measure earthquake intensity but PGA is measured by instruments, such as accelerographs. It can be correlated to macroseismic intensities on the Mercalli scale but these correlations are associated with large uncertainty.

The peak horizontal acceleration (PHA) is the most commonly used type of ground acceleration in engineering applications. It is often used within earthquake engineering (including seismic building codes) and it is commonly plotted on seismic hazard maps. In an earthquake, damage to buildings and infrastructure is related more closely to ground motion, of which PGA is a measure, rather than the magnitude of the earthquake itself. For moderate earthquakes, PGA is a reasonably good determinant of damage; in severe earthquakes, damage is more often correlated with peak ground velocity.

Earthquake energy is dispersed in waves from the hypocentre, causing ground movement omnidirectionally but typically modelled horizontally (in two directions) and vertically. PGA records the acceleration (rate of change of speed) of these movements, while peak ground velocity is the greatest speed (rate of movement) reached by the ground, and peak displacement is the distance moved. These values vary in different earthquakes, and in differing sites within one earthquake event, depending on a number of factors. These include the length of the fault, magnitude, the depth of the quake, the distance from the epicentre, the duration (length of the shake cycle), and the geology of the ground (subsurface). Shallow-focused earthquakes generate stronger shaking (acceleration) than intermediate and deep quakes, since the energy is released closer to the surface.

Peak ground acceleration can be expressed in g (the acceleration due to Earth's gravity, equivalent to g-force) as either a decimal or percentage; in m/s^2 (1 g = 9.81 m/s^2); or in Gal, where 1 Gal is equal to 0.01 m/s^2 (1 g = 981 Gal).

The ground type can significantly influence ground acceleration, so PGA values can display extreme variability over distances of a few kilometers, particularly with moderate to large earthquakes. The varying PGA results from an earthquake can be displayed on a shake map. Due to the complex conditions affecting PGA, earthquakes of similar magnitude can offer disparate results, with many moderate magnitude earthquakes generating significantly larger PGA values than larger magnitude quakes.

During an earthquake, ground acceleration is measured in three directions: vertically (V or UD, for up-down) and two perpendicular horizontal directions (H1 and H2), often north-south (NS) and east-west (EW). The peak acceleration in each of these directions is recorded, with the highest individual value often reported. Alternatively, a combined value for a given station can be noted. The peak horizontal ground acceleration (PHA or PHGA) can be reached by selecting the higher individual recording, taking the mean of the two values, or calculating a vector sum of the two components. A three-component value can also be reached, by taking the vertical component into consideration also.

In seismic engineering, the effective peak acceleration (EPA, the maximum ground acceleration to which a building responds) is often used, which tends to be ⅔ – ¾ the PGA.

Seismic Risk and Engineering

Study of geographic areas combined with an assessment of historical earthquakes allows geologists to determine seismic risk and to create seismic hazard maps, which show the likely PGA values to be experienced in a region during an earthquake, with a probability of exceedance (PE). Seismic engineers and government planning departments use these values to determine the appropriate earthquake loading for buildings in each zone, with key identified structures (such as hospitals, bridges, power plants) needing to survive the maximum considered earthquake (MCE).

Damage to buildings is related to both peak ground velocity (PGA) and the duration of the earthquake – the longer high-level shaking persists, the greater the likelihood of damage.

Comparison of Instrumental and Felt Intensity

Peak ground acceleration provides a measurement of instrumental intensity, that is, ground shaking recorded by seismic instruments. Other intensity scales measure felt intensity, based on eyewitness reports, felt shaking, and observed damage. There is correlation between these scales, but not always absolute agreement since experiences and damage can be affected by many other factors, including the quality of earthquake engineering.

Generally speaking,

- 0.001 g (0.01 m/s^2) – perceptible by people.
- 0.02 g (0.2 m/s^2) – people lose their balance.
- 0.50 g – very high; well-designed buildings can survive if the duration is short.

Correlation with the Mercalli Scale

The United States Geological Survey developed an Instrumental Intensity scale, which maps peak ground acceleration and peak ground velocity on an intensity scale similar to the felt Mercalli scale. These values are used to create shake maps by seismologists around the world.

Instrumental Intensity	Acceleration (g)	Velocity (cm/s)	Perceived shaking	Potential damage
I	< 0.0017	< 0.1	Not felt	None
II–III	0.0017 – 0.014	0.1 – 1.1	Weak	None
IV	0.014 – 0.039	1.1 – 3.4	Light	None
V	0.039 – 0.092	3.4 – 8.1	Moderate	Very light
VI	0.092 – 0.18	8.1 – 16	Strong	Light
VII	0.18 – 0.34	16 – 31	Very strong	Moderate
VIII	0.34 – 0.65	31 – 60	Severe	Moderate to heavy
IX	0.65 – 1.24	60 – 116	Violent	Heavy
X+	> 1.24	> 116	Extreme	Very heavy

Other Intensity Scales

In the 7-class Japan Meteorological Agency seismic intensity scale, the highest intensity, Shindo 7, covers accelerations greater than 4 m/s^2 (0.41 g).

ACCELEROGRAPH

An accelerograph can be referred to as a strong-motion instrument or seismograph, or simply an earthquake accelerometer. They are usually constructed as a self-contained box, which previously included a paper or film recorder (an analogue instrument) but now they often record directly on digital media and then the data is transmitted via the Internet.

Accelerographs are useful for when the earthquake ground motion is so strong that it causes the more sensitive seismometers to go off-scale. There is an entire science of strong ground motion, that is dedicated to studying the shaking in the vicinity of earthquakes (roughly within about 100 km of the fault rupture).

Accelerographs record the acceleration of the ground with respect to time. This recording is often called an accelerograms, strong-motion record or acceleration time-history. From this record strong-motion intensity measures (IMs, also called parameters) can be computed. The simplest of which is peak ground acceleration (PGA). Other IMs include Arias intensity, peak ground velocity (PGV), for which the accelerogram needs to be integrated once, peak ground displacement (PGD), for which double integration is required. Often a response spectrum is computed to show how the earthquake would affect structures of different natural frequencies or periods. These observations are useful to assess the seismic hazard of an area.

As well as their engineering applications, accelerograms are also useful for the study earthquakes from a scientific viewpoint. For example, accelerograms can be used to reconstruct the detailed history of rupture along a fault during an earthquake, which would not be possible with seismograms from standard instruments because they would be too far away to resolve the details. An example of an accelerograph array that was established to improve knowledge of near-source earthquake shaking as well as earthquake rupture propagation is the Parkfield Experiment, which involved a massive set of strong motion instrumentation.

Within the accelerograph, there is an arrangement of three accelerometer sensing heads. In recent low-cost instruments these are usually micro-machined (MEMS) chips that are sensitive to one direction. Thus constructed, the accelerometer can measure full motion of the device in three dimensions.

Unlike the continually recording seismometer, accelerometers nearly always work in a triggered mode. That means a level of acceleration must be set which starts the recording process. For analogue and older digital instruments this makes maintenance much more difficult without a direct Internet connection (or some other means of communication). Many trips have been made to accelerometers after a large earthquake, only to find that the memory was filled with extraneous noise, or the instrument was malfunctioning.

Accelerometers are used to monitor the response of structures to earthquakes. Analysis of these records along with the shaking recorded at base of the structure can improve building design, through earthquake engineering.

Accelerogram

Accelerograph is an instrument that records the ground acceleration during an earthquake. These recordings from the accelerograph are what are called accelerogram.

There are three transducers in the accelerograph from which three components of the motion are recorded. One transducer records the vertical motion while the two remaining transducers record the horizontal motions.

The earliest accelerographs in the world were developed in the 1930s; from then on the numbers accelerograms which were recorded throughout the world has increased. Thus, there are advances in earthquake engineering in most of the developed world.

Analog accelerographs were the first to be used which were recorded on films. This procedure was costly and time consuming as it required digitations. Advances in technology in the decades to follow resulted in the introduction of the first digital accelerographs in the late 1970s.

Digital accelerographs are easier to operate and the outputs from them are good representation of the earthquakes compared to analog accelerographs. In addition to this digital accelerographs do not require the digitations process and fewer errors.

Generation of Artificial Accelerograms

Artificial accelerogram are generated to match the target response spectrum. Groups of accelerograms are used to generate artificial accelerograms. Superposition of sinusoidal waves is used to generate artificial accelerograms. Computer programs such as SIMQKE are used to generate artificial records.

Selection Procedures

The selection of the appropriate accelerograms for the given structural design condition or analysis is necessary, as this forms the basis for realistic design of structures which resist earthquakes loads that may possibly occur at the specified site. Selection of accelerograms is usually based on earthquake scenarios such as magnitude, distance and site classifications.

The magnitude of an earthquake is important information as it is a measure of the size of an earthquake in relation to the amount of energy that is released during the earthquake. The duration and the frequent content of the accelerogram are dependent on the magnitude of the accelerograms. Since the duration of an earthquake is how long it takes for the fault area to rupture, earthquakes with large fault area have large magnitudes. Thus, the duration and the magnitude of an earthquake are related as that they both associate with the fault area.

Selection of accelerograms according to site classification requires a site which has similar geological characteristics to the site under consideration.

When selecting recorded accelerograms it is very important that the fault-site distance is used. Accelerograms which are recorded at near-fault distance have different characteristics from accelerograms which are recorded at a distance far from the fault-site. This results from the fact that rupture fault influences the characteristics of the accelerogram. Accelerogram which is recorded near the earthquake source is often influenced by rupture of fault effect, such as rupture directivity and fling step.

The energy content and the duration of the accelerogram are influenced by the type of rupture directivity which is present in the location the accelerogram is recorded at. For that reason in the selection process there should be a check for the type of rupture directivity present.

If the rupture directivity is a forward directivity the recorded accelerogram will have high amplitude and short duration. A characteristic of forward directivity is that the fault rupture velocity will be near as large as the velocity of the shear wave for the given site. Thus, there will be a wave front of the earthquake which arrives as a large pulse of motion at the start of the recording. This results in the structure being damaged. This occurs when the site is away from the epicentre.

On the other hand if the rupture directivity is a backward directivity which takes place when the accelerogram recorded on a site that is close to the epicentre, the recorded acclerogram will have small amplitude and long duration.

Another effect which is not often discussed is fling-step. This takes place when the ground that is subjected to the earthquake deforms statically and as a consequence the accelerogram will have unidirectional velocity pulse.

Considering the affects of the distance and directivity on the accelerogram in this manner makes sure that the most influential effects of the distance which are likely to affect the accelerogram are taken into account. For example if the forward directivity is not considered in the selection and the recording for the site had forward directivity the structure will be damaged in case an earthquake occurs.

Although most researchers have agreed for many years that the parameters that should be considered when selecting appropriate accelerogram for a given site are magnitude, distance and site classifications, with greater emphasis given to the magnitude and the distance. In contrast to this conservative view Baker and Cornell consider the effect of epsilon (the difference of Sa of the recorded accelerograms from the predicted Sa from the ground motion predicted equation) on the structural response to be greater than the effect of distance and magnitude. Thus, indicating that for selection e is the most significant parameter of the accelerogram. Since, e is a good indicator of the value of spectral acceleration on the spectrum if it is a peak value or not.

Scaling Procedures and Biased Response due to Scaling

Accelerograms are scaled on condition that the required intensity levels could not be achieved by the selection of a given accelerogram. This results from the fact that when selecting an accelerogram only limited characteristics are taken into account, such as the magnitude, distance, site classifications and at times *e*. Thus, scaling the selected accelerogram ensures that the characteristics which were not considered or could not be considered (since there is a limit to the recorded accelerograms from the data bank for a particular location) match that of the design spectrum.

There are numerous ways of scaling accelerograms which are considered in the literature. Scaling accelerograms are considered according to: Peak ground acceleration, peak ground velocity, Arias intensity (AI), Root-mean-square (RMS), spectral acceleration (Sa) and various parameters of the accelerogram. Most of scaling parameters are not of great use since some have more advantage in representation of the hazard than others. Out of all the scaling parameters the one which is often applied and recommended by codes is scaling to the spectral acceleration (Sa) which is an intensity measure.

Scaling factor is multiplied to the time history when the accelerogram is being scaled in terms of amplitude. The amplitude of the time history is scaled up or down so that it meets the requirements. In this method of scaling the frequent content of the accelerogram is not altered.

Scaling accelerograms by a factor which is not close to one is more likely to bring about unrealistic earthquake ground motion as a result of increased in energy while the duration and the frequent content of the accelerogram stay unadjusted. When the accelerogram is scaled the resulting time history should have the features which are expected of a real earthquake, since if there is a great difference between the two, the scaled time history will not be of any use as it is unrealistic. For that reason scaling accelerograms should not be done if accelerograms that meet the design spectrum could not be found.

In addition, another point which must also be considered is scaling the time axis of the accelerogram which is used to adjust the frequent content. This is not often carried out since it increases the duration and the energy content of the accelerogram. Thus impractical accelerogram is produced. Compared to scaling of the amplitude it is more difficult to scale the time axis of the accelerogram from the fact that there is greater care to be taken when doing time axis scaling.

Response is said to be bias if there is a systematic difference between the responses of a scaled record compared to unscaled record at the same intensity measure.

Research findings show whether the response of the structure is degree of biased of a randomly selected records for a given magnitude and distance, as well as scaling dependant on: the scaling factor, strength of the structure, the structure's fundamental period of vibration and also the magnitude and the distance of the accelerogram.

From this it is clear that the scaling factor which is used is not the only factor that contributes to the existence of the bias response after scaling.

The findings of the above research also show that the bias response can be avoided if records are chosen carefully to have spectral shapes which are near enough to the spectral shape of the target. One particularly striking aspect of this is that the magnitude and the distance of the accelerogram are not of significance under this condition.

Bias response of a structure can be reduced by selecting the records according to Conditional mean spectrum (CMS-*e*) or *e* instead of magnitude and distance. Given that the structural response is to a great extent affected by the spectral shape (which *e* is an indirect measure of).

Bias response of a structure can be avoided or reduced by selecting accelerograms which have similar spectral shape to the target spectral shape. In addition, it is stressed that when selecting accelerograms the main emphasis should not be on the magnitude or the distance as they do not influence the spectra shape.

Recorded and Artificial Accelerograms

There is more useful information about the earthquake for a particular site that is contained in a recorded accelerogram. These include the accelerogram characteristics such as amplitude, duration and frequency. As well as this, recorded accelerograms contain useful information on the factors affecting the accelerogram such as directivity, rupture mechanism and source to site distance.

In addition, recorded accelerograms have the characteristics which are expected of real earthquakes in contrast artificial accelerograms can sometimes have unrealistic characteristic which could result in impractical design, although there are technologies to limit this impracticalities this days. For that reason real accelerograms are often selected in the design of structures.

The number of accelerograms which are recorded across the world has increased over the past few years at the same time these databanks do not cover all the seismically active areas. The reason for this is that some of these places do not have accelerographs in place to record the earthquakes. Consequently, it is very difficult to find recorded accelerograms that have the necessary magnitude, distance and soil type for a site which is located in a region with few or no accelerographs to record the accelerograms. This is clearly a disadvantage of recorded accelerograms.

Meanwhile, the benefit of artificial accelerograms when compared with real accelerograms is that few records are needed to create artificial acclerogram which almost always matches the design spectra. On the other hand, unlike artificial records real accelerograms can be scaled to match the design spectrum without affecting the essential characteristics such as the frequency content.

In trying to examine the many aspects of real and artificial accelerograms it is clear that both have their advantages and disadvantages for that reason it is always up to the engineer to select whichever is most suited for the design. For that reason the decision reached for this project is to use real accelerograms as they present less serious limitations than artificial accelerograms.

References

- Earthquake-instrumentation, earthquakes-tectonics-hazard-and-risk-mitigation: intechopen.com, Retrieved 19 July, 2020
- Bolt, Bruce (August 2005), Earthquakes: 2006 Centennial Update – The 1906 Big One (Fifth ed.), W. H. Freeman and Company, p. 110, ISBN 978-0716775485
- https://www.seis-insight.eu/en/public-2/planetary-seismology/how-a-seismometer-works
- Serva, Leonello; Vittori, Eutizio; Comerci, Valerio; Esposito, Eliana; Guerrieri, Luca; Michetti, Alessandro Maria; Mohammadioun, Bagher; Mohammadioun, Georgianna C.; Porfido, Sabina; Tatevossian, Ruben E. (2016). "Earthquake Hazard and the Environmental Seismic Intensity (ESI) Scale". Pure and Applied Geophysics. 173 (5): 1479–1515. Bibcode:2016PApGe.173.1479S. doi:10.1007/s00024-015-1177-8
- How-a-seismometer-works, planetary-seismology: seis-insight.eu, Retrieved 5 May, 2020
- "Technical Details on the GSC's Internet Accelerograph". Archived from the original on 2008-03-11. Retrieved 2008-08-24
- Use-of-earthquake-accelerograms, engineering: ukessays.com, Retrieved 6 August, 2020

3

Seismology: A Comprehensive Study

Seismology refers to the study of earthquakes and seismic waves. Seismic gap, seismic migration, seismic noise, seismo-electromagnetics, paleoseismology, etc. are studied under its domain. All these aspects related to seismology have been carefully analyzed in this chapter.

SEISMOLOGY

Seismology is the science of earthquakes and studies the causes and effects from minute pulsations to the most catastrophic natural phenomenon inside Earth. The methods are classified into two divisions based on energy source of the seismic waves. Earthquake seismology is caused by natural shock waves of earthquakes and derives information on physical properties, composition, and the gross internal structure of Earth. Explosion seismology is the product of artificial blasts: (1) detonating a charge of dynamite (land) and (2) nonexplosive vibroseis or compressed air (marine) at selected sites to infer information about regional/local structures. This is extensively being applied to interpret the interfaces of rock boundaries, layered sedimentary sequences, location of water tables, and oil and gas exploration.

The survey works on the mode of propagation of waves in elastic media; more precisely, travel in rock media. The subsurface unit is assumed to be homogeneous and isotropic to simplify wave propagation resulting in basic interpretation of the measured effects at the plane of discontinuity.

Seismology contributes to tectonic and structural geology mostly through earthquake source studies or, more specifically, through being able to estimate the location, geometry and size of slip on faults. Earthquake seismology made a fundamental contribution to the discovery and confirmation of plate tectonics through studies that used fault plane solutions to demonstrate plate rigidity and the nature of plate boundaries. It is notable how quickly these studies appeared after the installation of the World Wide Standardized Seismograph Network (WWSSN) in the early 1960s made reliable determinations of focal mechanisms possible. In the 1970s earthquake seismology was responsible for important advances in understanding large-scale continental tectonics but had little influence on more detailed structural geology on the continents until about the 1980s, for two reasons, which are linked. Firstly, the continents deform in ways that are more complicated than the oceans, with earthquakes and faults distributed over wide areas and not confined to the narrow plate boundaries that typify the oceans. There are indeed patterns in the active faulting in such continental regions, but they were not easy to recognize with the relatively few earthquake focal mechanisms available soon after the installation of the WWSSN. Secondly, earthquake source seismology could not contribute much to debates in structural geology through the 1970s, because it was too inaccurate.

In particular, first motion fault plane solutions were rarely well-constrained in orientation, and routinely reported focal depths based on teleseismic arrival time data were notoriously unreliable, unable to tell whether an earthquake was in the crust or the mantle, let alone within the crystalline basement or sedimentary cover.

Two advances helped to increase the impact of seismology on structural geology by the late 1970s. The first was the installation of many dense local seismograph networks, often temporary and designed to monitor aftershocks, which greatly improved knowledge of focal depth distributions outside the few areas where permanent dense networks already existed. The second was the development of synthetic seismogram techniques that allowed much more accurate determination of earthquake source parameters from teleseismic waveforms themselves, rather than from just the arrival times and first motion polarities (see Appendix on the Handbook). For earthquakes larger than Mw ~ 5.5 it became possible, under favorable conditions, to estimate fault geometry (strike, dip, rake) to within 5–15°, centroid depth to within ~±3 km, and seismic moment to within 10–20%.

As more abundant and better quality earthquake data became available on the continents the natural advantages of seismology could then be exploited:

- Earthquake source seismology could provide information about the fault geometry underground which, when combined with surface observations, could give an overall three-dimensional image of the fault.

Faults could be studied as they moved, in their original orientation, unaffected by uplift, rotation or tilting.

Large areas could be investigated quickly to look for patterns of fault movement, in the sure knowledge that faults which are seismically active now are all part of the same deformational episode.

With these abilities earthquake seismology had something new to offer. Hitherto, structural and tectonic geology had necessarily been based on field or seismic reflection studies of faults that are now inactive. Many of these old structures are no longer in their original orientation, and demonstrating that old faults over a wide region had been active simultaneously was a major effort, if it was possible at all. Moreover, comparative studies between inactive regions were difficult, limited by the area any field geologist could reasonably address. Earthquake source studies have certainly not supplanted or made redundant the older field- and seismic reflection-based methods of structural investigation, each of which have their own obvious advantages. However, by offering insights that were simply unobtainable by other means, earthquake seismology has profoundly influenced thoughts and concepts in tectonic geology over the last 20 years.

Seismology is primarily an observational science; its scientists rarely have the opportunity to control the levels of variables (such as temperature, stress, or depth) that might influence their results. Consequently, classical design considerations, which dominate the subjects of agricultural trials or sample surveys, play a relatively minor role in most seismological experiments. This does not mean, however, that principles of design have no place in seismology. Some aspects of the experiments can be controlled: the number of observations or length of observation period; the quality and consistency of data collection; the division of time or space regions into parts for comparison; the need for replication and careful verification of results; avoidance of (or at the very least, making allowance for) the use of the same data to both set up the experiment (in terms of selection of subregions, etc.) and to vindicate its conclusions; and so on. All of these are vitally important aspects. While only common sense principles are required, attention to them is likely to be more effective in avoiding unsound conclusions than the use of the latest and most sophisticated techniques. As any statistical consultant will confirm, the most valuable time to consult a statistician is before the experiment commences, not after the data has been collected.

SEISMIC WAVE

Seismic waves are waves of energy that travel through the Earth's layers, and are a result of earthquakes, volcanic eruptions, magma movement, large landslides and large man-made explosions that give out low-frequency acoustic energy. Many other natural and anthropogenic sources create low-amplitude waves commonly referred to as ambient vibrations. Seismic waves are studied by geophysicists called seismologists. Seismic wave fields are recorded by a seismometer, hydrophone (in water), or accelerometer.

The propagation velocity of seismic waves depends on density and elasticity of the medium as well as the type of wave. Velocity tends to increase with depth through Earth's crust and mantle, but drops sharply going from the mantle to the outer core.

Earthquakes create distinct types of waves with different velocities; when reaching seismic observatories, their different travel times help scientists to locate the source of the hypocenter. In geophysics the refraction or reflection of seismic waves is used for research into the structure of the Earth's interior, and man-made vibrations are often generated to investigate shallow, subsurface structures.

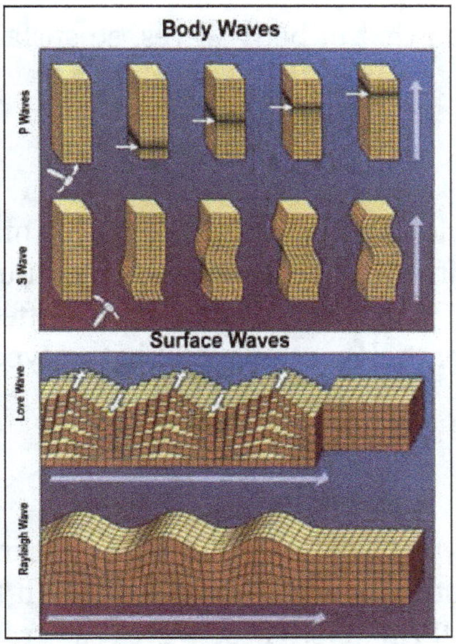

Body waves and surface waves.

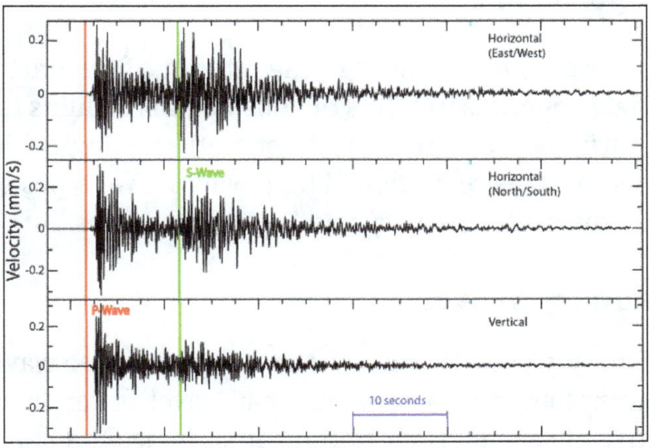

p-wave and s-wave from seismograph.

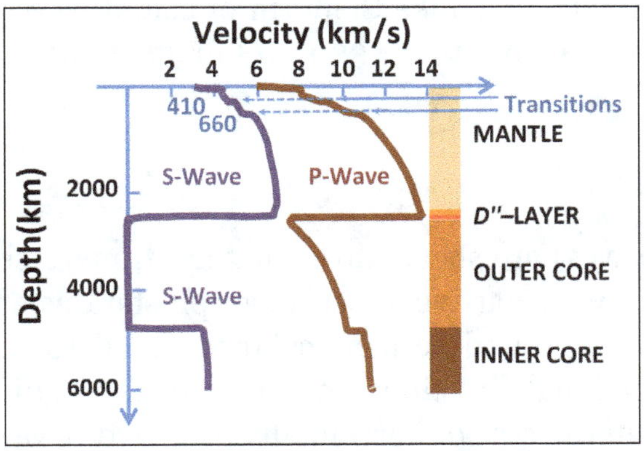

Velocity of seismic waves in the Earth versus depth. The negligible *S*-wave velocity in the outer core occurs because it is liquid, while in the solid inner core the *S*-wave velocity is non-zero.

Types

Among the many types of seismic waves, one can make a broad distinction between body waves, which travel through the Earth, and surface waves, which travel at the Earth's surface.

Other modes of wave propagation exist than those described in this article; though of comparatively minor importance for earth-borne waves, they are important in the case of asteroseismology.

- Body waves travel through the interior of the Earth.
- Surface waves travel across the surface. Surface waves decay more slowly with distance than body waves, which travel in three dimensions.
- Particle motion of surface waves is larger than that of body waves, so surface waves tend to cause more damage.

Body Waves

Body waves travel through the interior of the Earth along paths controlled by the material properties in terms of density and modulus (stiffness). The density and modulus, in turn, vary according to temperature, composition, and material phase. This effect resembles the refraction of light waves. Two types of particle motion result in two types of body waves: Primary and Secondary waves.

Primary Waves

Primary waves (P-waves) are compressional waves that are longitudinal in nature. P-waves are pressure waves that travel faster than other waves through the earth to arrive at seismograph stations first, hence the name "Primary". These waves can travel through any type of material, including fluids, and can travel nearly 1.7 times faster than the S-waves. In air, they take the form of sound waves, hence they travel at the speed of sound. Typical speeds are 330 m/s in air, 1450 m/s in water and about 5000 m/s in granite.

Secondary Waves

Secondary waves (S-waves) are shear waves that are transverse in nature. Following an earthquake event, S-waves arrive at seismograph stations after the faster-moving P-waves and displace the ground perpendicular to the direction of propagation. Depending on the propagational direction, the wave can take on different surface characteristics; for example, in the case of horizontally polarized S waves, the ground moves alternately to one side and then the other. S-waves can travel only through solids, as fluids (liquids and gases) do not support shear stresses. S-waves are slower than P-waves, and speeds are typically around 60% of that of P-waves in any given material. Shear waves can't travel through any liquid medium, so the absence of S-wave in earth's outer core suggests a liquid state.

Surface Waves

Seismic surface waves travel along the Earth's surface. They can be classified as a form of mechanical surface waves. They are called surface waves, as they diminish as they get further from the surface. They travel more slowly than seismic body waves (P and S). In large earthquakes, surface waves can have an amplitude of several centimeters.

Rayleigh Waves

Rayleigh waves, also called ground roll, are surface waves that travel as ripples with motions that are similar to those of waves on the surface of water (note, however, that the associated particle motion at shallow depths is retrograde, and that the restoring force in Rayleigh and in other seismic waves is elastic, not gravitational as for water waves). The existence of these waves was predicted by John William Strutt, Lord Rayleigh, in 1885. They are slower than body waves, roughly 90% of the velocity of S waves for typical homogeneous elastic media. In a layered medium (like the crust and upper mantle) the velocity of the Rayleigh waves depends on their frequency and wavelength.

Love Waves

Love waves are horizontally polarized shear waves (SH waves), existing only in the presence of a semi-infinite medium overlain by an upper layer of finite thickness. They are named after A.E.H. Love, a British mathematician who created a mathematical model of the waves in 1911. They usually travel slightly faster than Rayleigh waves, about 90% of the S wave velocity, and have the largest amplitude.

Stoneley Waves

A Stoneley wave is a type of boundary wave (or interface wave) that propagates along a solid-fluid boundary or, under specific conditions, also along a solid-solid boundary. Amplitudes of Stoneley waves have their maximum values at the boundary between the two contacting media and decay exponentially towards the depth of each of them. These waves can be generated along the walls of a fluid-filled borehole, being an important source of coherent noise in VSPs and making up the low frequency component of the source in sonic logging.

Normal Modes

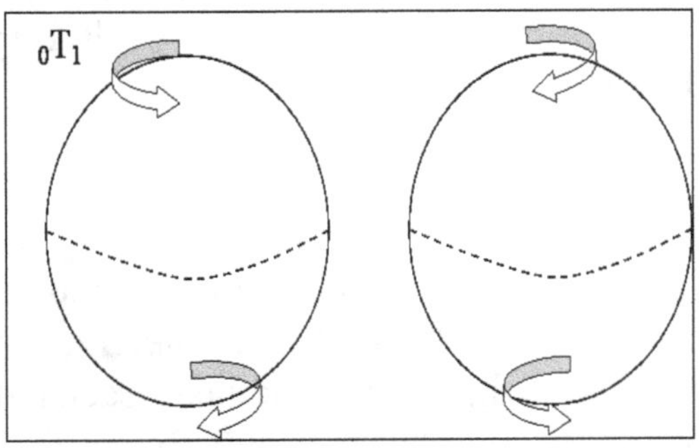

The sense of motion for toroidal $_0T_1$ oscillation for two moments of time.

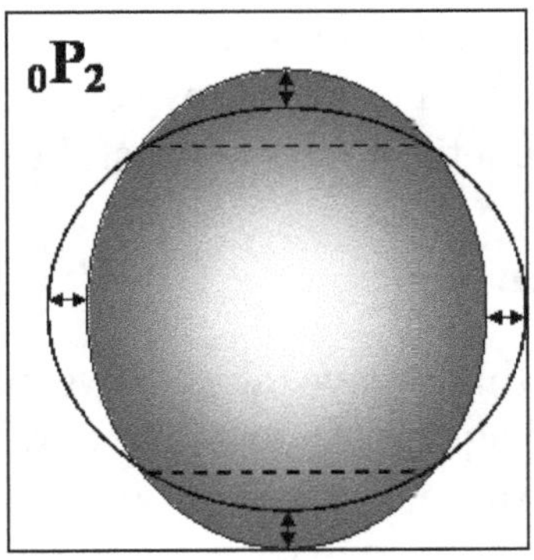

The scheme of motion for spheroidal $_0S_2$ oscillation. Dashed lines give nodal (zero) lines. Arrows give the sense of motion.

Free oscillations of the Earth are standing waves, the result of interference between two surface waves traveling in opposite directions. Interference of Rayleigh waves results in *spheroidal oscillation S* while interference of Love waves gives *toroidal oscillation T*. The modes of oscillations are specified by three numbers, e.g., ${}_nS_l^m$, where *l* is the angular order number (or *spherical harmonic degree*,). The number *m* is the azimuthal order number. It may take on 2*l*+1 values from -*l* to +*l*. The number *n* is the *radial order number*. It means the wave with *n* zero crossings in radius. For spherically symmetric Earth the period for given *n* and *l* does not depend on *m*.

Some examples of spheroidal oscillations are the "breathing" mode ${}_0S_0$, which involves an expansion and contraction of the whole Earth, and has a period of about 20 minutes; and the "rugby" mode ${}_0S_2$, which involves expansions along two alternating directions, and has a period of about 54 minutes. The mode ${}_0S_1$ does not exist because it would require a change in the center of gravity, which would require an external force.

Of the fundamental toroidal modes, ${}_0T_1$ represents changes in Earth's rotation rate; although this occurs, it is much too slow to be useful in seismology. The mode ${}_0T_2$ describes a twisting of the northern and southern hemispheres relative to each other; it has a period of about 44 minutes.

The first observations of free oscillations of the Earth were done during the great 1960 earthquake in Chile. Presently periods of thousands of modes are known. These data are used for determining some large scale structures of the Earth interior.

P and S Waves in Earth's Mantle and Core

When an earthquake occurs, seismographs near the epicenter are able to record both P and S waves, but those at a greater distance no longer detect the high frequencies of the first S wave. Since shear waves cannot pass through liquids, this phenomenon was original evidence for the now well-established observation that the Earth has a liquid outer core, as demonstrated by Richard Dixon Oldham. This kind of observation has also been used to argue, by seismic testing, that the Moon has a solid core, although recent geodetic studies suggest the core is still molten.

Notation

The path that a wave takes between the focus and the observation point is often drawn as a ray diagram. An example of this is shown in a figure above. When reflections are taken into account there are an infinite number of paths that a wave can take. Each path is denoted by a set of letters that describe the trajectory and phase through the Earth. In general an upper case denotes a transmitted wave and a lower case denotes a reflected wave. The two exceptions to this seem to be "g" and "n".

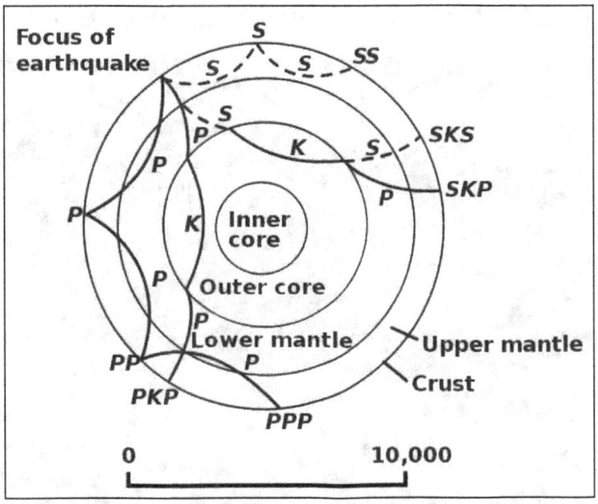

Earthquake wave paths.

c	The wave reflects off the outer core
d	A wave that has been reflected off a discontinuity at depth d
g	A wave that only travels through the crust
i	A wave that reflects off the inner core
I	A P-wave in the inner core
h	A reflection off a discontinuity in the inner core
J	An S wave in the inner core
K	A P-wave in the outer core
L	A Love wave sometimes called LT-Wave (Both caps, while an Lt is different)
n	A wave that travels along the boundary between the crust and mantle
P	A P wave in the mantle
p	A P wave ascending to the surface from the focus
R	A Rayleigh wave
S	An S wave in the mantle
s	An S wave ascending to the surface from the focus
w	The wave reflects off the bottom of the ocean
	No letter is used when the wave reflects off of the surfaces

For example:

- ScP is a wave that begins traveling towards the center of the Earth as an S wave. Upon reaching the outer core the wave reflects as a P wave.
- sPKIKP is a wave path that begins traveling towards the surface as an S-wave. At the surface it reflects as a P-wave. The P-wave then travels through the outer core, the inner core, the outer core, and the mantle.

Usefulness of P and S Waves in Locating an Event

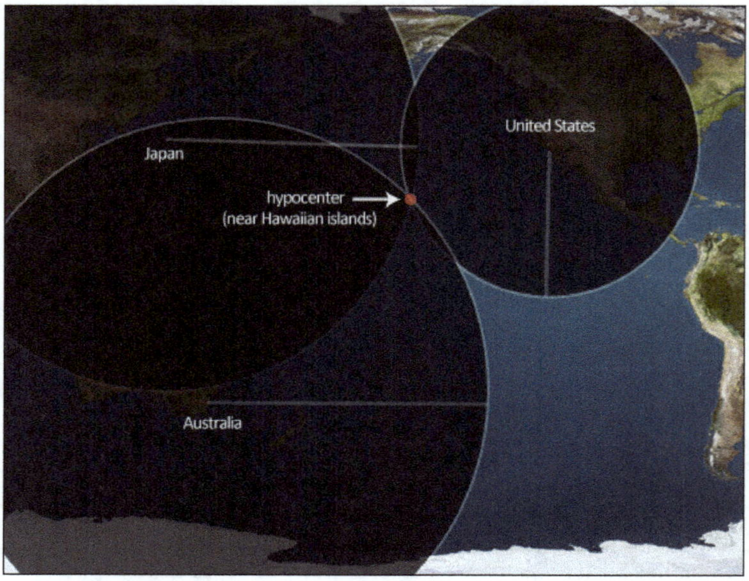

The Hypocenter/Epicenter of an earthquake is calculated by using the seismic data of that earthquake from at least three different locations.

In the case of local or nearby earthquakes, the difference in the arrival times of the P and S waves can be used to determine the distance to the event. In the case of earthquakes that have occurred at global distances, three or more geographically diverse observing stations (using a common clock) recording P-wave arrivals permits the computation of a unique time and location on the planet for the event. Typically, dozens or even hundreds of P-wave arrivals are used to calculate hypocenters. The misfit generated by a hypocenter calculation is known as "the residual". Residuals of 0.5 second or less are typical for distant events, residuals of 0.1–0.2 s typical for local events, meaning most reported P arrivals fit the computed hypocenter that well. Typically a location program will start by assuming the event occurred at a depth of about 33 km; then it minimizes the residual by adjusting depth. Most events occur at depths shallower than about 40 km, but some occur as deep as 700 km.

A quick way to determine the distance from a location to the origin of a seismic wave less than 200 km away is to take the difference in arrival time of the P wave and the S wave in seconds and multiply by 8 kilometers per second. Modern seismic arrays use more complicated earthquake location techniques.

At teleseismic distances, the first arriving P waves have necessarily travelled deep into the mantle, and perhaps have even refracted into the outer core of the planet, before travelling back up to the Earth's surface where the seismographic stations are located. The waves travel more quickly than if they had traveled in a straight line from the earthquake. This is due to the appreciably increased velocities within the planet, and is termed Huygens' Principle. Density in the planet increases with depth, which would slow the waves, but the modulus of the rock increases much more, so deeper means faster. Therefore, a longer route can take a shorter time.

The travel time must be calculated very accurately in order to compute a precise hypocenter. Since P waves move at many kilometers per second, being off on travel-time calculation by even a half second can mean an error of many kilometers in terms of distance. In practice, P arrivals from many stations are used and the errors cancel out, so the computed epicenter is likely to be quite accurate, on the order of 10–50 km or so around the world. Dense arrays of nearby sensors such as those that exist in California can provide accuracy of roughly a kilometer, and much greater accuracy is possible when timing is measured directly by cross-correlation of seismogram waveforms.

P-wave

A P-wave is one of the two main types of elastic body waves, called seismic waves in seismology. P-waves travel faster than other seismic waves and hence are the first signal from an earthquake to arrive at any affected location or at a seismograph. P-waves may be transmitted through gases, liquids, or solids.

The name P-wave can stand for either pressure wave (as it is formed from alternating compressions and rarefactions) or primary wave (as it has high velocity and is therefore the first wave to be recorded by a seismograph).

Seismic Waves in the Earth

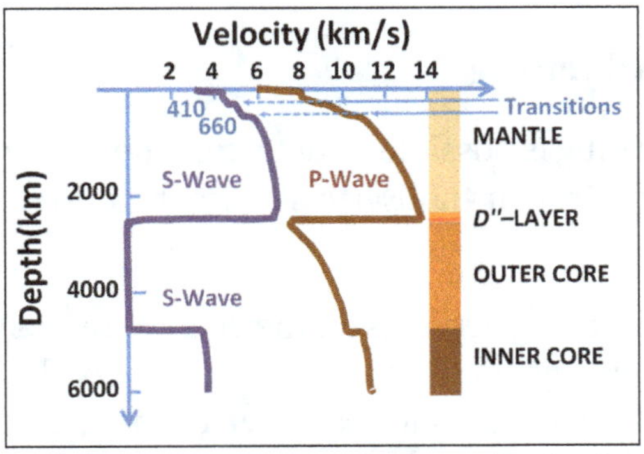

Velocity of seismic waves in the Earth versus depth. The negligible *S*-wave velocity in the outer core occurs because it is liquid, while in the solid inner core the *S*-wave velocity is non-zero.

Primary and secondary waves are body waves that travel within the Earth. The motion and behavior of both *P*-type and *S*-type in the Earth are monitored to probe the interior structure of the Earth. Discontinuities in velocity as a function of depth are indicative of changes in phase or composition. Differences in arrival times of waves originating in a seismic event like an earthquake as a result of waves taking different paths allow mapping of the Earth's inner structure.

P-wave Shadow Zone

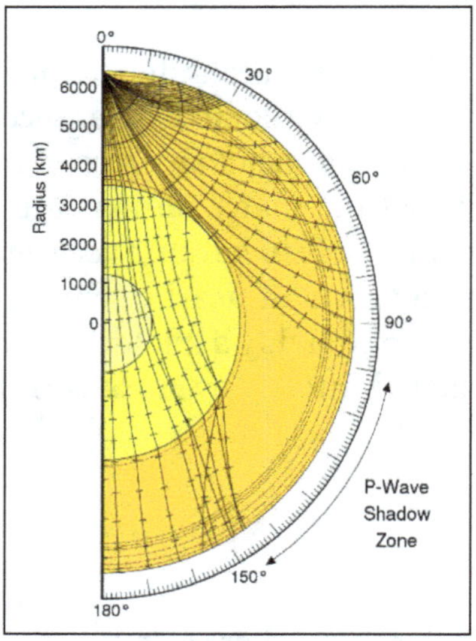

P-wave shadow zone.

Almost all the information available on the structure of the Earth's deep interior is derived from observations of the travel times, reflections, refractions and phase transitions of seismic body waves, or normal modes. P-waves travel through the fluid layers of the Earth's interior, and yet they are refracted slightly when they pass through the transition between the semisolid mantle and the liquid outer core. As a result, there is a P-wave "shadow zone" between 103° and 142° from the earthquake's focus, where the initial P-waves are not registered on seismometers. In contrast, S-waves do not travel through liquids.

As an Earthquake Warning

Advance earthquake warning is possible by detecting the nondestructive primary waves that travel more quickly through the Earth's crust than do the destructive secondary and Rayleigh waves.

The amount of advance warning depends on the delay between the arrival of the P-wave and other destructive waves, generally on the order of seconds up to about 60 to 90 seconds for deep, distant, large quakes such as the 2011 Tohoku earthquake. The effectiveness of advance warning depends on accurate detection of the P-waves and rejection of ground vibrations caused by local activity (such as trucks or construction). Earthquake early warning systems can be automated to allow for immediate safety actions, such as issuing alerts, stopping elevators at the nearest floors and switching off utilities.

Propagation

Velocity

In isotropic and homogeneous solids, a P-wave travels in a straight line longitudinal;

thus, the particles in the solid vibrate along the axis of propagation (the direction of motion) of the wave energy. The velocity of P-waves in such a medium is given by

$$v_p = \sqrt{\frac{K + \frac{4}{3}\mu}{\rho}} = \sqrt{\frac{\lambda + 2\mu}{\rho}}$$

where K is the bulk modulus (the modulus of incompressibility), μ is the shear modulus (modulus of rigidity, sometimes denoted as G and also called the second Lamé parameter), ρ is the density of the material through which the wave propagates, and λ is the first Lamé parameter.

In typical situations the interior of the Earth, the density ρ usually varies much less than K or μ, so the velocity is mostly "controlled" by these two parameters.

The elastic moduli P-wave modulus, M, is defined so that $M = K + 4\mu / 3$ and thereby,

$$v = \sqrt{M / \rho}$$

Typical values for P-wave velocity in earthquakes are in the range 5 to 8 km/s. The precise speed varies according to the region of the Earth's interior, from less than 6 km/s in the Earth's crust to 13.5 km/s in the lower mantle, and 11 km/s through the inner core.

Table: Velocity of common rock types.

Rock Type	Velocity [m/s]	Velocity [ft/s]
Unconsolidated Sandstone	4600 - 5200	15000 - 17000
Consolidated Sandstone	5800	19000
Shale	1800 - 4900	6000 -16000
Limestone	5800 - 6400	19000 - 21000
Dolomite	6400 - 7300	21000 - 24000
Anhydrite	6100	20000
Granite	5800 - 6100	19000 - 20000
Gabbro	7200	23600

Geologist Francis Birch discovered a relationship between the velocity of P waves and the density of the material the waves are traveling in:

$$V_p = a(\bar{M}) + b\rho$$

which later became known as Birch's law.

S-wave

In seismology, S-waves, secondary waves, or shear waves (sometimes called an elastic S-wave) are a type of elastic wave, and are one of the two main types of elastic body waves, so named because they move through the body of an object, unlike surface waves.

The S-wave is a transverse wave, meaning that, in the simplest situation, the oscillations of the particles of the medium is perpendicular to the direction of wave propagation, and the main restoring force comes from shear stress.

Its name, S for secondary, comes from the fact that it is the second direct arrival on an earthquake seismogram, after the compressional primary wave, or P-wave, because S-waves travel slower in rock. Unlike the P-wave, the S-wave cannot travel through the molten outer core of the Earth, and this causes a shadow zone for S-waves opposite to where they originate. They can still appear in the solid inner core: when a P-wave strikes the boundary of molten and solid cores at an oblique angle, S-waves will form and propagate in the solid medium. When these S-waves hit the boundary again at an oblique angle they will in turn create P-waves that propagate through the liquid medium. This property allows seismologists to determine some physical properties of the inner core.

Isotropic Medium

For the purpose of this explanation, a solid medium is considered isotropic if its strain (deformation) in response to stress is the same in all directions. Let $u = (u_1, u_2, u_3)$ be the displacement vector of a particle of such a medium from its "resting" position $x = (x_1, x_2, x_3)$ due elastic vibrations, understood to be a function of the rest position x and time t. The deformation of the medium at that point can be described by the strain tensor e, the 3×3 matrix whose elements are:

$$e_{ij} = \frac{1}{2}(\partial_i u_j + \partial_j u_i)$$

where ∂_i denotes partial derivative with respect to position coordinate x_i. The strain tensor is related to the 3×3 stress tensor τ by the equation,

$$\tau_{ij} = \lambda \delta_{ij} \sum_k e_{kk} + 2\mu e_{ij}$$

Here δ_{ij} is the Kronecker delta (1 if $i = j$, 0 otherwise) and λ and μ are the Lamé parameters (μ being the material's shear modulus). It follows that:

$$\tau_{ij} = \lambda \delta_{ij} \left(\sum_k \partial_k u_k\right) + \mu(\partial_i u_j + \partial_j u_i)$$

From Newton's law of inertia, one also gets,

$$\rho \partial_t^2 u_i = \sum_j \partial_j \tau_{ij}$$

where ρ is the density (mass per unit volume) of the medium at that point, and ∂_t denotes partial derivative with respect to time.

Combining the last two equations one gets the seismic wave equation in homogeneous media,

$$\rho\partial_t^2 u_i = \lambda\partial_i\left(\sum_k \partial_k u_k\right) + \mu\left(\sum_j \partial_i\partial_j u_j + \mu\partial_j\partial_j u_i\right)$$

Using the nabla operator notation of vector calculus, $\nabla = (\partial_1, \partial_2, \partial_3)$, with some approximations, this equation can be written as,

$$\rho\partial_t^2 u = (\lambda + 2\mu)\nabla(\nabla \cdot \boldsymbol{u}) - \mu\nabla \times (\nabla \times \boldsymbol{u})$$

Taking the curl of this equation and applying vector identities, one gets,

$$\partial_t^2(\nabla \times \boldsymbol{u}) = \frac{\mu}{\rho}\nabla^2(\nabla \times \boldsymbol{u})$$

This formula is the wave equation applied to the vector quantity $\nabla \times \boldsymbol{u}$, which is the material's shear strain. Its solutions, the S-waves, are linear combinations of sinusoidal plane waves of various wavelengths and directions of propagation, but all with the same speed $\beta = \sqrt{\mu / \rho}$.

Taking the divergence of seismic wave equation in homogeneous media, instead of the curl, yields a wave equation describing propagation of the quantity $\nabla \cdot \boldsymbol{u}$, which is the material's compression strain. The solutions of this equation, the P-waves, travel at the speed $\alpha = \sqrt{(\lambda + 2\mu) / \rho}$ that is more than twice the speed β of S-waves.

The steady-state SH waves are defined by the Helmholtz equation:

$$(\nabla^2 + k^2)\boldsymbol{u} = 0$$

SEISMIC GAP

A seismic gap is a segment of an active fault known to produce significant earthquakes that has not slipped in an unusually long time, compared with other segments along the same structure. There is a hypothesis or theory that states that over long periods of time, the displacement on any segment must be equal to that experienced by all the other parts of the fault. Any large and longstanding gap is, therefore, considered to be the fault segment most likely to suffer future earthquakes.

The applicability of this approach has been criticised by some seismologists although earthquakes sometimes have occurred in previously-identified seismic gaps.

Loma Prieta Seismic Gap, California: Prior to the 1989 Loma Prieta earthquake, that segment of the San Andreas fault system recorded much less seismic activity than other parts of the fault. The main shock and aftershocks of the 1989 event occurred within the previous seismic gap.

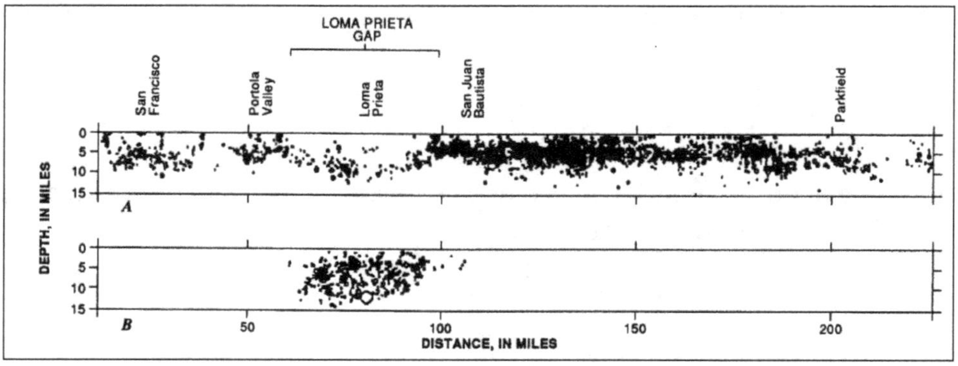

Cross sections along the San Andreas fault showing recorded seismic activity A) 20 years before the Loma Prieta event, B) The main shock (large circle) and aftershocks for the Loma Prieta event, USGS Circular 1045.

Central Kuril Gap, Russia: Immediately following the 2004 Indian Ocean earthquake, a seismic gap analysis of the seismic zones around the Pacific Ocean identified the Central Kuril segment of the Kuril-Kamchatka Trench subduction zone as the most likely to give rise to a major earthquake. This zone, 500 km in length, at that time had experienced no major earthquake since 1780, but was bounded to north and south by segments that had moved within the last 100 years. The M_w = 8.3 earthquake of 15 November 2006 and the M_w = 8.2 earthquake of 13 January 2007 occurred within the defined gap.

Central Himalayan Gap, India: Although there had been earthquakes to the west (near Delhi) in 1905, and to the east (Nepal–Bihar earthquake) in 1934, there was a 600-kilometer-long region of the central Himalayan that had not ruptured since 1505. In April 2015, the 7.8 M_w April 2015 Nepal earthquake occurred near the center of this region.

SEISMIC MIGRATION

Seismic migration is the process by which seismic events are geometrically re-located in either space or time to the location the event occurred in the subsurface rather than the location that it was recorded at the surface, thereby creating a more accurate image of the subsurface. This process is necessary to overcome the limitations of geophysical methods imposed by areas of complex geology, such as: faults, salt bodies, folding, etc.

Migration moves dipping reflectors to their true subsurface positions and collapses diffractions, resulting in a migrated image that typically has an increased spatial resolution and resolves areas of complex geology much better than non-migrated images. A form of migration is one of the standard data processing techniques for reflection-based geophysical methods (seismic reflection and ground-penetrating radar).

The need for migration has been understood since the beginnings of seismic exploration and the very first seismic reflection data from 1921 were migrated. Computational migration algorithms have been around for many years but they have only entered wide usage in the past 20 years because they are extremely resource-intensive. Migration can lead to a dramatic uplift in image quality so algorithms are the subject of intense research, both within the geophysical industry as well as academic circles.

Rationale

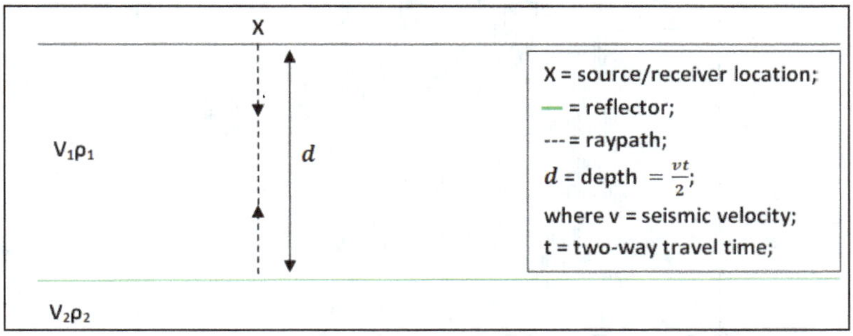

The raypath for a zero-offset reflection from a horizontal reflector.

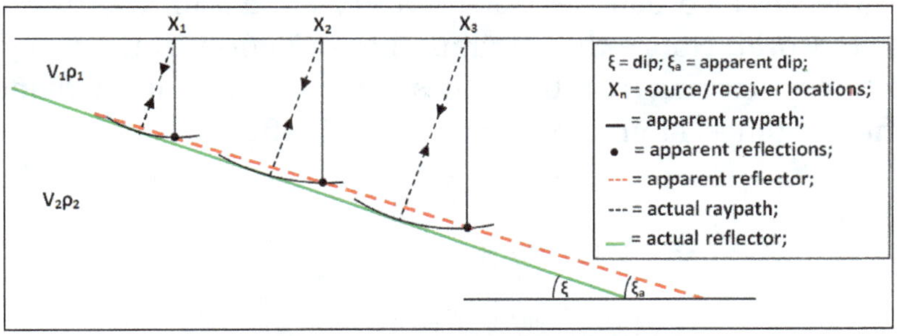

The raypath for a zero-offset reflection from a dipping reflector and the resultant apparent dip.

A zero-offset migrated data set. This data was migrated using a time-migration referred to as phase-shift which operates in the Fourier domain. The migration has replaced all events in their correct locations, successfully reconstructing a syncline. However, there are erroneous events (swinging arcs) throughout the image which are migration induced noise.

Seismic waves are elastic waves that propagate through the Earth with a finite velocity, governed by the elastic properties of the rock in which they are travelling. At an interface between two rock types, with different acoustic impedances, the seismic energy is either refracted, reflected back towards the surface or attenuated by the medium. The reflected energy arrives at the surface and is recorded by geophones that are placed at a known distance away from the source of the waves. When a geophysicist views the recorded energy from the geophone, they know both the travel time and the distance between the source and the receiver, but not the distance down to the reflector.

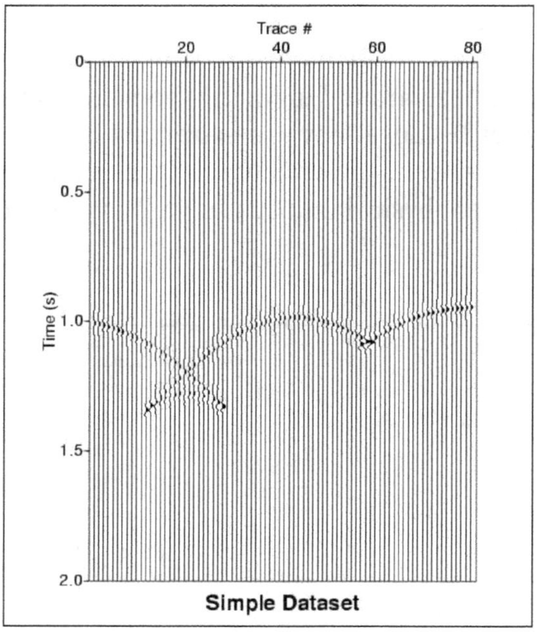

A zero-offset non-migrated data set. Raw zero-offset data for a simple syncline in a constant velocity world. Notice the signature bow-tie effect in the image. This is the result of reflections occurring from both sides of the syncline, and arriving at the same receiver at different times. Migration can correct this effect.

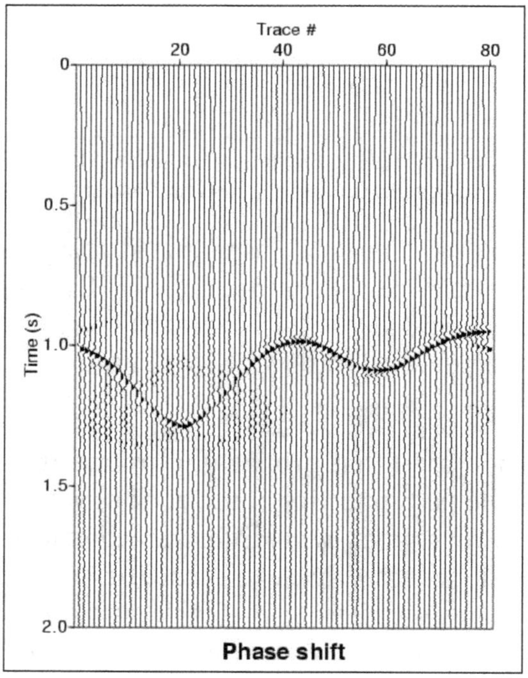

In the simplest geological setting, with a single horizontal reflector, a constant velocity and a source and receiver at the same location (referred to as zero-offset, where offset is the distance between the source and receiver), the geophysicist can determine the location of the reflection event by using the relationship:

$$d = \frac{vt}{2},$$

where d is the distance, v is the seismic velocity (or rate of travel) and t is the measured time from the source to the receiver.

In this case, the distance is halved because it can be assumed that it only took one-half of the total travel time to reach the reflector from the source, then the other half to return to the receiver.

The result gives us a single scalar value, which actually represents a half-sphere of distances, from the source/receiver, which the reflection could have originated from. It is a half-sphere, and not a full sphere, because we can ignore all possibilities that occur above the surface as unreasonable. In the simple case of a horizontal reflector, it can be assumed that the reflection is located vertically below the source/receiver point.

The situation is more complex in the case of a dipping reflector, as the first reflection originates from further up the direction of dip and therefore the travel-time plot will show a reduced dip that is defined the "migrator's equation":

$$\tan \xi_a = \sin \xi,$$

where ξ_a is the *apparent dip* and ξ is the true dip.

Zero-offset data is important to a geophysicist because the migration operation is much simpler, and can be represented by spherical surfaces. When data is acquired at non-zero offsets, the sphere becomes an ellipsoid and is much more complex to represent (both geometrically, as well as computationally).

Use

For a geophysicist, complex geology is defined as anywhere there is an abrupt or sharp contrast in lateral and/or vertical velocity (e.g. a sudden change in rock type or lithology which causes a sharp change in seismic wave velocity).

Some examples of what a geophysicist considers complex geology are: faulting, folding, (some) fracturing, salt bodies, and unconformities. In these situations a form of migration is used called pre-stack migration (PreSM), in which all traces are migrated before being moved to zero-offset. Consequently, much more information is used, which results in a much better image, along with the fact that PreSM honours velocity changes more accurately than post-stack migration.

Types of Migration

Depending on budget, time restrictions and the subsurface geology, geophysicists can employ 1 of 2 fundamental types of migration algorithms, defined by the domain in which they are applied: time migration and depth migration.

Time Migration

Time migration is applied to seismic data in time coordinates. This type of migration makes the assumption of only mild lateral velocity variations and this breaks down in the presence of most interesting and complex subsurface structures, particularly salt. Some popularly used time migration algorithms are: Stolt migration, Gazdag and Finite-difference migration.

Depth Migration

Depth Migration is applied to seismic data in depth (regular Cartesian) coordinates, which must be calculated from seismic data in time coordinates. This method does therefore require a velocity model, making it resource-intensive because building a seismic velocity model is a long and iterative process. The significant advantage to this migration method is that it can be successfully used in areas with lateral velocity variations, which tend to be the areas that are most interesting to petroleum geologists. Some of the popularly used depth migration algorithms are Kirchhoff depth migration, Reverse Time Migration (RTM), Gaussian Beam Migration and Wave-equation migration.

Resolution

The goal of migration is to ultimately increase spatial resolution and one of the basic assumptions made about the seismic data is that it only shows primary reflections and all noise has been removed. In order to ensure maximum resolution (and therefore maximum uplift in image quality) the data should be sufficiently pre-processed before migration. Noise that may be easy to distinguish pre-migration could be smeared across the entire aperture length during migration, reducing image sharpness and clarity.

A further basic consideration is whether to use 2D or 3D migration. If the seismic data has an element of cross-dip (a layer that dips perpendicular to the line of acquisition) then the primary reflection will originate from out-of-plane and 2D migration cannot put the energy back to its origin. In this case, 3D migration is needed to attain the best possible image.

Modern seismic processing computers are more capable of performing 3D migration, so the question of whether to allocate resources to performing 3D migration is less of a concern.

Graphical Migration

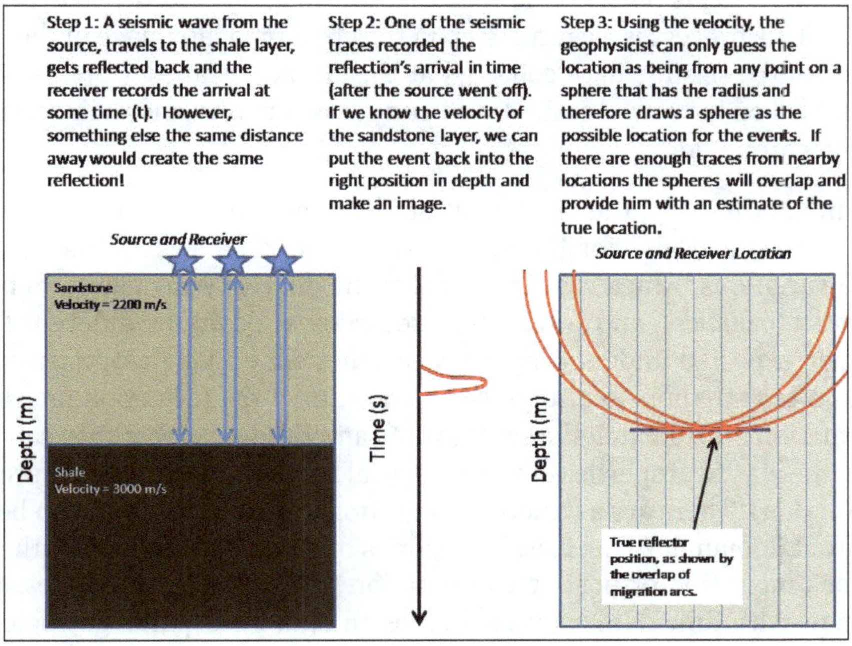

An example of simple graphical migration. Until the advent of modern computers in the 1960s and 1970s this was a method used by geophysicists to primitively 'migrate' their data. This method is obsolete with the advent of digital processors, but is useful for understanding the basic principle behind migration.

The simplest form of migration is that of graphical migration. Graphical migration assumes a constant velocity world and zero-offset data, in which a geophysicist draws spheres or circles from the receiver to the event location for all events. The intersection of the circles then form the reflector's "true" location in time or space. An example of such can be seen in the diagram.

Technical Details

Migration of seismic data is the correction of the flat-geological-layer assumption by a numerical, grid-based spatial convolution of the seismic data to account for dipping events (where geological layers are not flat). There are many approaches, such as the popular Kirchhoff migration, but it is generally accepted that processing large spatial sections (apertures) of the data at a time introduces fewer errors, and that depth migration is far superior to time migration with large dips and with complex salt bodies.

Basically, it repositions/moves the energy (seismic data) from the recorded locations to the locations with the correct common midpoint (CMP). While the seismic data is received at the proper locations originally (according to the laws of nature), these locations do not correspond with the assumed CMP for that location. Though stacking the data without the migration corrections yields a somewhat inaccurate picture of the subsurface, migration is preferred for better most imaging recorder to drill and maintain oilfields. This process is a central step in the creation of an image of the subsurface from active source seismic data collected at the surface, seabed, boreholes, etc., and therefore is used on industrial scales by oil and gas companies and their service providers on digital computers.

Explained in another way, this process attempts to account for wave dispersion from dipping reflectors and also for the spatial and directional seismic wave speed (heterogeneity) variations, which cause wavefields (modelled by ray paths) to bend, wave fronts to cross (caustics), and waves to be recorded at positions different from those that would be expected under straight ray or other simplifying assumptions. Finally, this process often attempts to also preserve and extract the formation interface reflectivity information imbedded in the seismic data amplitudes, so that they can be used to reconstruct the elastic properties of the geological formations (amplitude preservation, seismic inversion). There are a variety of migration algorithms, which can be classified by their output domain into the broad categories of time migration or depth migration, and pre-stack migration or post-stack migration (orthogonal) techniques. Depth migration begins with time data converted to depth data by a spatial geological velocity profile. Post-stack migration begins with seismic data which has already been stacked, and thus already lost valuable velocity analysis information.

SEISMIC NOISE

Seismic noise is a generic name for a relatively persistent vibration of the ground, due to a multitude of causes, that is a non-interpretable or unwanted component of signals recorded by seismometers.

Physically, seismic noise consists mostly of surface waves. Low frequency waves (below 1 Hz) are generally called microseisms; high frequency waves (above 1 Hz) are called microtremors. Its causes include nearby human activities (such as traffic or heavy machinery), winds and other atmospheric phenomena, and ocean waves.

Seismic noise is relevant to any discipline that depends on seismology, such as geology, oil exploration, hydrology, and earthquake engineering, and structural health monitoring. It is often called ambient wavefield or ambient vibrations in those disciplines. (However, the latter term may also refer to vibrations transmitted through by air, building, or supporting structures.)

Seismic noise is a nuisance for activities that are sensitive to vibrations, such as accurate measurements, precision milling, telescopes, and crystal growing. On the other hand, seismic noise does have some practical uses, for example to determine the low-strain dynamic properties of civil-engineering structures, such as bridges, buildings, and dams; or to determine the elastic properties of the soil and subsoil in order to draw seismic microzonation maps showing the predicted ground response to earthquakes.

Causes

Research on the origin of seismic noise indicates that the low frequency part of the spectrum (below 1 Hz) is due to natural causes, chiefly ocean waves. In particular the peak between 0.1 and 0.3 Hz is clearly associated with the interaction of water waves of nearly equal frequencies but opposite directions. At high frequency (above 1 Hz), seismic noise is mainly produced by human activities such as road traffic and industrial work; but there are also natural sources, like rivers. Around 1 Hz, wind and other atmospheric phenomena are also a major source of ground vibrations.

Physical Characteristics

The amplitude of seismic noise vibrations is typically in the order of 0.1 to 10 μm/s. High and low noise models as a function of frequency have been proposed.

The seismic noise includes a small number of body waves (P- and S-waves), but surface waves (Love and Rayleigh waves) predominate. These waves are dispersive, meaning that their phase velocity varies with frequency (generally, it decreases with increasing frequency). Since the dispersion curve (phase velocity or slowness as a function of frequency) is tightly related to the variations of the shear-wave velocity with depth in the different ground layers, it can be used as a non-invasive tool to investigate the underground structure.

Applications to Civil Engineering

After the 1933 Long Beach earthquake in California, a large experiment campaign led by D. S. Carder in 1935 allowed to record and analyze ambient vibrations in more than 200 buildings. These data were used in the design codes to estimate resonance frequencies of buildings but the interest of the method went down until the 1950s. Interest on ambient vibrations in structures grew further, especially in California and Japan, thanks to the work of earthquake engineers, including G. Housner, D. Hudson, K. Kanai, T. Tanaka, and others.

Ambient vibrations were however supplanted - at least for some time - by forced vibration techniques that allow to increase the amplitudes and control the shaking source and their system identification methods. Even though M. Trifunac showed in 1972 that ambient and forced vibrations led to the same results, the interest in ambient vibration techniques only rose in the late 1990s. They have now become quite attractive, due to their relatively low cost and convenience, and to the recent improvements in recording equipment and computation methods. The results of their low-strain dynamic probing were shown to be close enough to the dynamic characteristics measured under strong shaking, at least as long as the buildings are not severely damaged.

Scientific Study and Applications in Geology

The recording of seismic noise directly from the ground started in the 1950s with the enhancement of seismometers to monitor nuclear tests and the development of seismic arrays. The main contributions at that time for the analysis of these recordings came from the Japanese seismologist K. Aki in 1957. He proposed several methods used today for local seismic evaluation, such as Spatial Autocorrelation (SPAC), Frequency-wavenumber (FK), and correlation. However, the practical implementation of these methods was not possible at that time because of the low precision of clocks in seismic stations.

Again, improvements in instrumentation and algorithms led to renewed interest on those methods in the 1990s. Y.Nakamura rediscovered in 1989 the Horizontal to Vertical Spectral Ratio (H/V) method to derive the resonance frequency of sites. Assuming that shear waves dominate the microtremor, Nakamura observed that the H/V spectral ratio of ambient vibrations was roughly equal to the S-wave transfer function between the ground surface and the bedrock at a site.

In the late 1990s, array methods applied to seismic noise data started to yield ground properties in terms of shear waves velocity profiles. The European Research project SESAME worked to standardize the use of seismic noise to estimate the amplification of earthquakes by local ground characteristics.

Current use of Ambient Vibrations

Characterization of the Ground Properties

The analysis of the ambient vibrations leads to different products used to characterize the ground properties. From the easiest to the most complicated, these products are: power spectra, H/V peak, dispersion curves and autocorrelation functions.

Single-station Methods:

- Computation of power spectra, e.g. Passive seismic.
- HVSR (H/V spectral ratio): The H/V technique is especially related to ambient vibration recordings. Bonnefoy-Claudet et al. showed that peaks in the horizontal to vertical spectral ratios can be linked to the Rayleigh ellipticity peak, the Airy phase of the Love waves and/or the SH resonance frequencies depending on the proportion of these different types of waves in the ambient noise. By chance, all these values give however approximately the same value for a given ground so that H/V peak is a reliable method to estimate the resonance frequency of the sites. For 1 sediment layer on the bedrock, this value f_0 is related to the

velocity of S-waves V_s and the depth of the sediments H following: $f_0 = \frac{V_s}{4H}$. It can therefore be used to map the bedrock depth knowing the S-wave velocity. This frequency peak allows to constrain the possible models obtain using other seismic methods but is not enough to derive a complete ground model. Moreover, it has been shown that the amplitude of the H/V peak was not related to the magnitude of the amplification.

Array Methods: Using an array of seismic sensors recording simultaneously the ambient vibrations allow to understand more deeply the wavefield and therefore to derive more properties of the ground. Due to the limitation of the available number of sensors, several arrays of different sizes may be realized and the results merged. The information of the Vertical components is only linked to the Rayleigh waves, and therefore easier to interpret, but method using the 3 space components are also developed, providing information about Rayleigh and Love wavefield.

- FK, HRFK using the Beamforming technique.
- SPAC (Spatial Auto-correlation) method.
- Correlations methods.
- Refraction microtremor ReMI.

Characterization of the Vibration Properties of Civil Engineering Structures

Like earthquakes, ambient vibrations force into vibrations the civil engineering structures like bridges, buildings or dams. This vibration source is supposed by the greatest part of the used methods to be a white noise, i.e. with a flat noise spectrum so that the recorded system response is actually characteristic of the system itself. The vibrations are perceptible by humans only in rare cases (bridges, high buildings). Ambient vibrations of buildings are also caused by wind and internal sources (machines, pedestrians) but these sources are generally not used to characterize structures. The branch that studies the modal properties of systems under ambient vibrations is called Operational modal analysis (OMA) or Output-only modal analysis and provides many useful methods for civil engineering. The observed vibration properties of structures integrate all the complexity of these structures including the load-bearing system, heavy and stiff non-structural elements (infill masonry panels), light non-structural elements (windows) and the interaction with the soil (the building foundation may not be perfectly fixed on the ground and differential motions may happen). This is emphasized because it is difficult to produce models able to be compared with these measurements.

Single-station Methods: The power spectrum computation of ambient vibration recordings in a structure (e.g. at the top floor of a building for larger amplitudes) gives an estimation of its resonance frequencies and eventually its damping ratio.

Transfer Function Method: Assuming ground ambient vibrations is the excitation source of a structure, for instance a building, the Transfer Function between the bottom and the top allows to remove the effects of a non-white input. This may particularly be useful for low signal-to-noise ratio signals (small building/high level of ground vibrations). However this method generally is not able to remove the effect of soil-structure interaction.

Arrays: They consist in the simultaneous recording in several points of a structure. The objective is to obtain the modal parameters of structures: resonance frequencies, damping ratios and modal shapes for the whole structure. Notice than without knowing the input loading, the participation factors of these modes cannot a priori be retrieved. Using a common reference sensor, results for different arrays can be merged.

- Methods based on correlations: Several methods use the power spectral density matrices of simultaneous recordings, i.e. the cross-correlation matrices of these recordings in the Fourier domain. They allow to extract the operational modal parameters (Peak Picking method) that can be the results of modes coupling or the system modal parameters (Frequency Domain Decomposition method).

- System identification methods: Numerous system identification methods exist in the literature to extract the system properties and can be applied to ambient vibrations in structures

Inversion/Model Updating/Multi-model Approach

The obtained results cannot directly give information on the physical parameters (S-wave velocity, structural stiffness) of the ground structures or civil engineering structures. Therefore, models are needed to compute these products (dispersion curve, modal shapes) that could be compared with the experimental data. Computing a lot of models to find which agree with the data is solving the Inverse problem. The main issue of inversion is to well explore the parameter space with a limited number of computations of the model. However, the model fitting best the data is not the most interesting because parameter compensation, uncertainties on both models and data make many models with different input parameters as good compared to the data. The sensitivity of the parameters may also be very different depending on the model used. The inversion process is generally the weak point of these ambient vibration methods.

Material Needed

The acquisition chain is mainly made of a seismic sensor and a digitizer. The number of seismic stations depends on the method, from single point (spectrum, HVSR) to arrays (3 sensors and more). Three components (3C) sensors are used except in particular applications. The sensor sensitivity and corner frequency depend also on the application.

For ground measurements, velocimeters are necessary since the amplitudes are generally lower than the accelerometers sensitivity, especially at low frequency. Their corner frequency depends on the frequency range of interest but corner frequencies lower than 0.2 Hz are generally used. Geophones (generally 4.5 Hz corner frequency or greater) are generally not suited. For measurements in civil engineering structures, the amplitude is generally higher as well as the frequencies of interest, allowing the use of accelerometers or velocimeters with a higher corner frequency. However, since recording points on the ground may also be of interest in such experiments, sensitive instruments may be needed. Except for single station measurements, a common time stamping is necessary for all the stations. This can be achieved by GPS clock, common start signal using a remote control or the use of a single digitizer allowing the recording of several sensors. The relative location of the recording points is needed more or less precisely for the different techniques, requiring either manual distance measurements or differential GPS location.

Advantages and Limitations

The advantages of ambient vibration techniques compared to active techniques commonly used in exploration geophysics or earthquake recordings used in Seismic tomography:

- Relatively cheap, non-invasive and non-destructive method.
- Applicable to urban environment.
- Provide valuable information with little data (e.g. HVSR).
- Dispersion curve of Rayleigh wave relatively easy to retrieve.
- Provide reliable estimates of Vs30.

Limitations of these methods are linked to the noise wavefield but especially to common assumptions made in seismic:

- Penetration depth depends on the array size but also on the noise quality, resolution and aliasing limits depend on the array geometry.
- Complexity of the wavefield (Rayleigh, Love waves, interpretation of higher modes).
- Plane wave assumption for most of the array methods (problem of sources within the array).
- 1D assumption of the underground structure, even though 2D was also undertaken.
- Inverse problem difficult to solve as for many geophysical methods.

SEISMO-ELECTROMAGNETICS

Seismo-electromagnetics are various electro-magnetic phenomena believed to be generated by tectonic forces acting on the earth's crust, and possibly associated with seismic activity such as earthquakes and volcanoes. Study of these has been prompted by the prospect they might be generated by the increased stress leading up to an earthquake, and might thereby provide a basis for short-term earthquake prediction. However, despite many studies, no form of seismo-electromagnetics has been shown to be effective for earthquake prediction. A key problem is that earthquakes themselves produce relatively weak electromagnetic phenomena, and the effects from any precursory phenomena are likely to be too weak to measure. Close monitoring of the Parkfield earthquake revealed no significant pre-seismic electromagnetic effects. However, some researchers remain optimistic, and searches for seismo-electromagnetic earthquake precursors continue.

VAN Method

The VAN method – named after P. Varotsos, K. Alexopoulos and K. Nomicos, authors of the 1981 papers describing it – measures low frequency electric signals, termed "seismic electric signals" (SES), by which Varotsos and several colleagues claimed to have successfully predicted earthquakes in Greece. Both the method itself and the manner by which successful predictions were claimed have been severely criticized and debated by VAN, but the critics have not retracted their views.

Since 2001, the VAN group has introduced a concept they call "natural time", applied to the analysis of their precursors. Initially it is applied on SES to distinguish them from noise and relate them to a possible impending earthquake. In case of verification (classification as "SES activity"), natural time analysis is additionally applied to the general subsequent seismicity of the area associated with the SES activity, in order to improve the time parameter of the prediction. The method treats earthquake onset as a critical phenomenon.

After 2006, VAN say that all alarms related to SES activity have been made public by posting at arxiv.org. One such report was posted on Feb. 1, 2008, two weeks before the largest earthquake in Greece during the period 1983–2011. This earthquake occurred on February 14, 2008, with magnitude (M_w) 6.9. VAN's report was also described in an article in the newspaper Ethnos on Feb. 10, 2008. However, Gerassimos Papadopolous complained that the VAN reports were confusing and ambiguous, and that "none of the claims for successful VAN predictions is justified."

QuakeFinder and 'Freund Physics'

In his investigations of crystalline physics, Friedemann Freund found that water molecules embedded in rock can dissociate into ions if the rock is under intense stress. The resulting charge carriers can generate battery currents under certain conditions. Freund suggested that perhaps these currents could be responsible for earthquake precursors such as electromagnetic radiation, earthquake lights and disturbances of the plasma in the ionosphere. The study of such currents and interactions is known as "Freund physics".

Most seismologists reject Freund's suggestion that stress-generated signals can be detected and put to use as precursors, for a number of reasons. First, it is believed that stress does not accumulate rapidly before a major earthquake, and thus there is no reason to expect large currents to be rapidly generated. Secondly, seismologists have extensively searched for statistically reliable electrical precursors, using sophisticated instrumentation, and have not identified any such precursors. And thirdly, water in the earth's crust would cause any generated currents to be absorbed before reaching the surface.

QuakeFinder is a company focused on developing a system for earthquake prediction. The company has a long-standing collaboration with Freund. They have deployed a network of sensor stations that detect the electromagnetic pulses the team believes precede major earthquakes. Each sensor is believed to have a range of approximately 10 miles (16 km) from the instrument to the source of the pulses. As of 2016, the company says they have 125 stations in California, and their affiliate Jorge Heraud says he has 10 sites in Peru. Using these sensors, Heraud says that he has been able to triangulate pulses seen from multiple sites, in order to determine the origin of the pulses. He said that the pulses are seen beginning from 11 to 18 days before an impending earthquake, and have been used to determine the location and timing of future seismic events.

However, insofar as a verifiable prediction would require a publicly-stated announcement of the location, time, and size of an impending event before its occurrence, neither Quakefinder nor Heraud have yet verifiably predicted an earthquake, much less issued multiple predictions of the type that might be objectively testable for statistical significance.

Corralitos Anomaly

In the month prior to the 1989 Loma Prieta earthquake measurements of the earth's magnetic field at ultra-low frequencies by a magnetometer in Corralitos, California, just 7 km from the epicenter of the impending earthquake, started showing anomalous increases in amplitude. Just three hours before the quake the measurements soared to about thirty times greater than normal, with amplitudes tapering off after the quake. Such amplitudes had not been seen in two years of operation, nor in a similar instrument located 54 km away. To many people such apparent locality in time and space suggested an association with the earthquake.

Additional magnetometers were subsequently deployed across northern and southern California, but after ten years, and several large earthquakes, similar signals have not been observed. More recent studies have cast doubt on the connection, attributing the Corralitos signals to either unrelated magnetic disturbance or, even more simply, to sensor-system malfunction.

Study of the closely monitored 2004 Parkfield earthquake found no evidence of precursory electromagnetic signals of any type.

ULF Magnetic Field Precursors

Two recent studies by Konstantinos Eftaxias and his colleagues examined ULF magnetic fields preceding major earthquakes. At the 2011 Tohoku earthquake, ULF radiation exhibited critical behavior, while at the 2008 Sichuan earthquake, the researchers found a depression of the horizontal ULF magnetic field, which may also be interpreted as a manifestation of criticality.

TEC Variations

Professor Kosuke Heki of Hokkaido University in Japan said that he discovered by accident that GPS signals changed about 40 minutes before the 2011 Tohoku-Oki earthquake. Reviewing historical data for other earthquakes, he found that this same correlation occurred during other incidents. He suggested that the GPS signals were detecting variations in the levels of the TEC (total electron content) of the ionosphere in the hour preceding an earthquake.

Satellite Observations

The "Detection of Electro-Magnetic Emissions Transmitted from Earthquake Regions" satellite, constructed by CNES, has made observations which show strong correlations between certain types of low frequency electromagnetic activity and the most seismically active zones on the Earth, and have shown a sharp signal in the ionospheric electron density and temperature near southern Japan seven days before a 7.1 magnitude earthquake occurred there.

Quakesat is an earth observation nanosatellite based on 3 CubeSats. It was designed to be a proof-of-concept for collecting extremely low frequency earthquake precursor signals from space. The primary instrument is a magnetometer housed in a 2-foot (0.6 m) telescoping boom.The science behind the concept is disputed.

ESPERIA is an equatorial space mission mainly concerned with detecting any tectonic and preseismic related signals. More in general, it has been proposed for defining the near-Earth electromagnetic, plasma, and particle environment, and for studying perturbations and instabilities in the ionosphere-magnetosphere transition region. To study earthquake preparation processes and anthropogenic impacts in the Earth's surface, a phase A study has been realized for the Italian Space Agency.

The Deformation, Ecosystem Structure and Dynamics of Ice (DESDynI) radar satellite, which was canceled in the White House's 2012 budget proposal, would have the capacity to identify elastic strain in tectonic plates, combining L-band interferometric synthetic aperture radar and a multi-beam infrared lidar to detect strains in the Earth's surface that could lead to serious earthquakes.

Russia and the United Kingdom agreed to jointly deploy two satellites in 2015 that will measure electromagnetic signals that are released from the earth's crust prior to earthquakes. The project is said to be able to "help predict earthquakes and potentially save thousands of lives."

Another site of current research is China, where a satellite launch was planned for 2014, to provide data from ionospheric phenomena for comparison with seismo-electromagnetic phenomena on the ground. Such a link is partially borne out in the current literature, with ionospheric phenomena already shown to precede seismic phenomena by a few hours to days. The network would potentially show whether such ionospheric phenomena are sourced from ground electrical phenomena.

PALEOSEISMOLOGY

Fault scarp produced during the
South Napa earthquake in the Napa fault.

Earth scientists can gather data at key sites along sections of a fault to figure out the past timeline of earthquakes at each spot. This type of research is called paleoseismology.

The basic assumption that paleoseismologists use is that what happened in the past will most likely happen again in the future. Therefore, the more information that can be gathered about the past, the better idea we'll have about what is likely to happen in the future. This is important because we need to be able to design bridges and big buildings to withstand future large earthquakes, and to plan so we can be resilient in the days and years after the big earthquakes occur. This means we need paleoseismologists to collect information about when past earthquakes occurred on faults and how large the earthquakes were.

When it comes to sleuthing for evidence of past earthquakes, there is primary evidence and secondary evidence. Primary evidence includes clues that are created by the movement on the fault during the earthquake such as fault scarps, offset or folded layers of sediment and soil, and parts of the landscape that have been tilted, uplifted, down-dropped, or torn apart. The picture above shows an example of a fault scarp from the relatively small 2014 South Napa earthquake (M6.0). In larger earthquakes, the movement may be bigger. For example, a historic account from the 1857 M7.9 earthquake on the San Andreas Fault in southern California, stated that "This rent was in some places 5-10 yards wide, the earth at times filling it up like ploughed furrows; at others the ground stood apart, leaving a deep fissure." Secondary evidence includes clues produced by the shaking during the earthquake such as sand blows, rockfalls, landslides, and damaged trees.

Finding Paleoearthquakes

Finding these paleoearthquakes requires going to the right place. First, scientists study the landscape to find places where the faults are active. Second, they must go to places where evidence for past earthquakes could be preserved – places where the soil and sediment will cover the fault scarp and preserve it for hundreds to thousands of years. For the most part, paleoseismologists only see evidence for the larger earthquakes, above about a M6, because below that magnitude they are too small to leave a mark on the landscape that is likely to be preserved. Consequently, paleoseismology mostly provides data on the biggest earthquakes with the potential to cause the most damage.

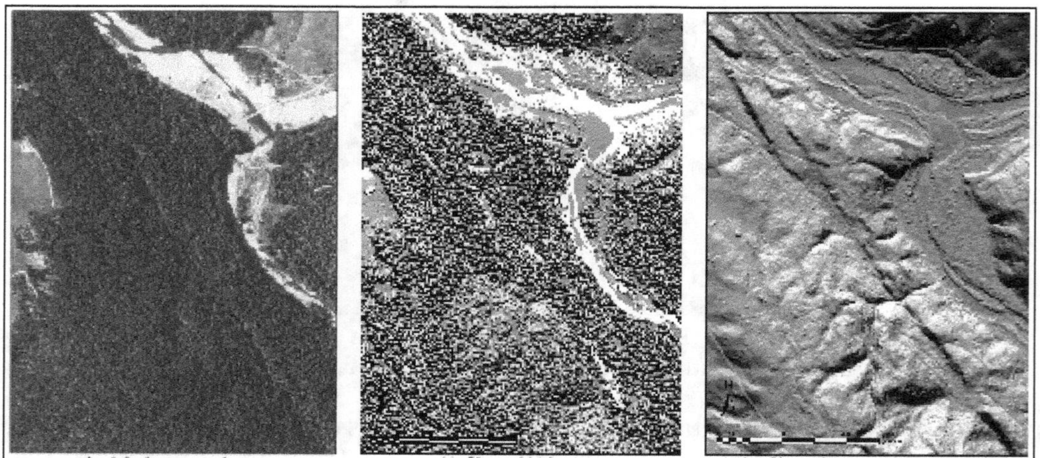
Aerial photograph Unfiltered Lidar Lidar filtered to remove vegetation

These three photos show on the left a photo from the air, in the middle a first return lidar, and on the right a bare earth lidar for a location along the San Andreas Fault where it crosses Annapolis Rd south of Gualala. The fault is difficult to see in the air photo and first return lidar, but is obvious in the bare earth lidar.

To find direct evidence for a paleoearthquake, first you need to find a fault. The use of lidar in geological investigations has made it much easier to map active faults. Lidar can cover a lot more ground than a few geologists hiking around looking for evidence of past earthquakes, and its ability to "see through" the vegetation also allows coverage in areas where it may be difficult or impossible for scientists to access.

Once you've found a fault, you need to be able to see the layers of sediment and soils that have been deposited over time beneath the surface. Usually this requires digging trenches across the fault. In the trench you can see sediments and soils that have come down from nearby hills and are deposited in layers, one on top of the other, like a stack of books. When an earthquake occurs along a fault, the layers are disrupted, bent and offset. As time goes by, new layers are deposited on top. Since layers deposited after the earthquake are not broken, it is clear that the earthquake occurred before they were deposited.

This pattern repeats over the centuries, creating a geological story in the layers and how the sediments are disrupted. Viewing a cross-section of deposits that cross the fault from the surface downward is like reading a timeline backwards from present to past. The job of a paleoseismologist is to unravel the sequence of events revealed in the cross-section.

Dead forest of western red cedar sticking up through a brackish-water tidal marsh at Copalis River on the Washington coast. The forest was killed during rapid subsidence accompanying the most recent Cascadia Subduction Zone earthquake. Tree rings from the dead trees provide evidence that the trees died sometime after the end of the A.D. 1699 growing season and before the beginning of the A.D. 1700 growing season. This age is consistent with the date of this earthquake based on a tsunami that was recorded in Japanese historical records. The forest in the background is above the highest tides.

In addition to trenching, there are some other rather creative techniques for identifying paleoearthquakes. Using dendrochronology, or tree rings, is an interesting way to look for paleoseismic evidence. Trees that were severely shaken, permanently tilted from shaking, or were inundated by salt water from subsidence, will suffer in health and not grow as quickly as usual for a while after an earthquake, or in some instances, the trees could be killed as a result of a large earthquake. The subsequent tree ring for that time period will be thinner than usual. In coastal areas like where this photo was taken, dendrochronology can be used to determine when large paleoearthquakes altered the level of the land: during subduction zone earthquakes some parts of the land surface can lower by several feet. Thus forests with trees going along a coast will suddenly die as the roots of the trees are drowned in salt water.

Determining the Age of a Paleoearthquake

How do you go from the observation of a prehistoric earthquake in the trench wall to the age of an earthquake? You'll want this information to understand how long ago the most recent paleoearthquake happened, how old the earthquake before that one is, and so on through the sequence of disrupted sediments.

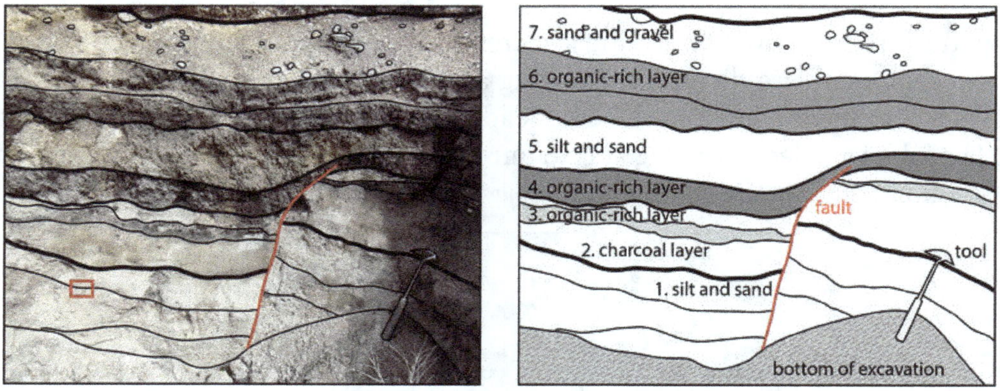

Fault produced during large magnitude earthquake on the San Andreas Fault at Pallett Creek, California when organic rich layer 4 was at the ground surface. Note that silt and sand layer 5 is thicker on the left side of the fault, where it filled in the step (scarp) made by motion on the fault. Layers 5 and 6 are not affected by the fault.

There are many different ways that dates are extracted from samples gathered at paleoearthquake sites. The dating technique that's familiar to most people, and is the most common, is radiocarbon dating. Radiocarbon dating allows you to determine the age of material that contains carbon, which is almost anything that was once a living plant or animal. Plant remains like seeds and leaves, charcoal from past fires, and shells and soils are all datable materials using radiocarbon.

Piece of charcoal in silt and sand layer 1 used in radiocarbon dating.

The use of accelerator mass spectrometry (AMS), which allows you to measure the number of radiogenic carbon isotopes in a sample and determine the age, has lead to improvements in paleoseismology. The big advantage is that new lab techniques can resolve the age of very small samples. This means that a scientist can collect samples the size of a grain of rice from a geologic feature instead of schlepping large, heavy samples from the field to the lab, and the data are even better, as a wide variety of material can be dated. The age of each earthquake is bracketed by dating layers above and below the youngest sediments deformed by the prehistoric earthquake.

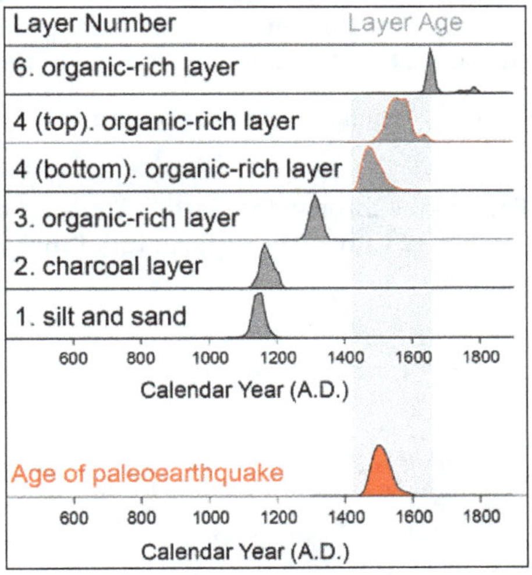

Radiocarbon ages are determined for all of the layers to establish that the dating is robust and no contamination has occurred. Note that samples get older with each deeper layer. The earthquake in the panel above occured when layer 4 was at the ground surface. The age of the earthquake is determined from the radiocarbon ages above and below, in this case the top and bottom of layer 4. This earthquake occured between 1457 and 1568 AD.

Estimating Paleoearthquake Magnitudes

Earthquake magnitudes are a reflection of the area of the fault rupture (the length and depth of the fault plane) and the amount of slip, or offset, that occurred on the fault. To estimate the magnitude of prehistoric earthquakes, scientists try to determine the amount of offset to features on the surface such as streams or gullies. The size of the offset is essentially correlated to the size of the earthquake. For example, an M7 will cause the ground to be offset by an average of about 1 meter, while a M7.5 has an average offset of about 2.6 meters. Secondary evidence may also be used to estimate the size of a region that was shaken by the earthquake. For example, the locations of sand blows and other liquefaction features or evidence of earthquake-induced landslides can be mapped across a region to determine the extent and severity of shaking. The pattern of shaking is then compared with the pattern of earthquakes of known magnitudes to determine what magnitude best fits the observations.

References

- F. Ardhuin, E. Stutzmann, M. Schimmel, and A. Mangeney (2011), Ocean wave sources of seismic noise. Journal of Geophysics Research, volume 115
- Seismology, earth-and-planetary-sciences: sciencedirect.com, Retrieved 26 August, 2020
- G. R. Helffrich & B. J. Wood (2002). "The Earth's mantle" (PDF). Nature. Macmillan Magazines. 412 (2 August): 501; Figure 1. doi:10.1038/35087500. PMID 11484043. Archived (PDF) from the original on 24 August 2016
- Chen, Yangkang; Chen, Hanming; Xiang, Kui; Chen, Xiaohong (2017). "Preserving the discontinuities in least-squares reverse time migration of simultaneous-source data". Geophysics. 82 (3): S185–S196. Bibcode:2017Geop...82S.185C. doi:10.1190/geo2016-0456.1
- Introduction-paleoseismology, science, earthquake-hazards, natural-hazards: usgs.gov, Retrieved 23 May, 2020
- CMR Fowler (2005). "§4.1 Waves through the Earth". The solid earth: an introduction to global geophysics (2nd ed.). Cambridge University Press. p. 100. ISBN 978-0-521-58409-8

4

Seismic Hazard Assessment

Seismic hazard assessment refers to the effort for prediction of potential ground shaking intensity from future earthquakes. It uses a common approach of probabilistic seismic hazard analysis in which the expected ground motions are evaluated at a given site. This chapter discusses the subject of seismic hazard assessment in detail.

Seismic hazard and risk are two very important concepts in engineering design and other policy considerations. Although seismic hazard and risk have often been used interchangeably, they are fundamentally different. Furthermore, seismic risk is more important in engineering design and other policy considerations. Seismic hazard assessment is an effort by earth scientists to quantify seismic hazard and its associated uncertainty in time and space and to provide seismic hazard estimates for seismic risk assessment and other applications. Although seismic hazard assessment is more a scientific issue, it deserves special attention because of its significant implication to society. Two approaches, probabilistic seismic hazard analysis (PSHA) and deterministic seismic hazard analysis (DSHA), are commonly used for seismic hazard assessment. Although PSHA has been proclaimed as the best approach for seismic hazard assessment, it is scientifically flawed (i.e., the physics and mathematics that PSHA is based on are not valid). Use of PSHA could lead to either unsafe or overly conservative engineering design or public policy, each of which has dire consequences to society. On the other hand, DSHA is a viable approach for seismic hazard assessment even though it has been labeled as unreliable. The biggest drawback of DSHA is that the temporal characteristics (i.e., earthquake frequency of occurrence and the associated uncertainty) are often neglected. An alternative, seismic hazard analysis (SHA), utilizes earthquake science and statistics directly and provides a seismic hazard estimate that can be readily used for seismic risk assessment and other applications.

Seismic hazard studies have become increasingly more important for earthquake engineering applications all around the world. It is possible to mitigate the damages of earthquake for the constructions by using engineering techniques. Seismic hazard assessment is commonly used to define and classify the susceptible areas and in the preparation of seismic zonation maps. The main purpose of the seismic hazard mapping is to obtain ground motion distribution in any place. The ground motion parameters include peak ground acceleration, peak velocity, peak displacement, and response spectral values or histories of acceleration, velocity and displacement gained from past earthquakes.

SEISMIC HAZARD ANALYSIS

Seismic hazard analysis involves the quantitative estimation of ground shaking hazards at a particular area. Seismic hazards can be analyzed deterministically as and when a particular earthquake scenario is assumed, or probabilistically, in which uncertainties in earthquake size, location, and time of occurrence are explicitly considered. A critical part of seismic hazard analysis is the determination of Peak Ground Acceleration (PGA) and response acceleration (spectral acceleration) for an area/site. Spectral acceleration (Sa) is preferred for the design of civil engineering structures. It is an accepted trend in engineering practice to develop design response spectrum for different types of foundation materials such as rock, hard soil and weak soils. Analysis of lineaments and faults helps in understanding the regional seismotectonic activity of the area. Lineaments are linear features seen on the surface of earth which represents faults, features, shear zones, joints, litho contacts, dykes, etc; and are of great relevance to geoscientists.

Deterministic Seismic Hazard Analysis

Deterministic seismic hazard analysis (DSHA) is an approach for evaluating site-specific seismic hazard that is influenced by the maximum hazard from the controlling sources affecting the specific study site. In its conventional form, DSHA does not consider sources other than the largest "controlling" source and it does not account for the time factors owing to the uncertainty of earthquakes occurrences in time. Under certain condition, ignoring these factors can lower the conservatism of the hazard estimate, especially when other non-controlling sources generate hazards nearly equivalent to that of the controlling source or when the structure's design life is longer than the controlling source earthquake's return period.

Deterministic seismic hazard involve a quantitative estimation of various ground shaking hazard for a particular area. DSHA is a simple process that is Useful especially where tectonic features are practically dynamic and sound defined. The spotlight is generally on determining the maximum credible earthquake (MCE) motion at the site. DSHA is founded on tectonic features, it tends to be conventional since the highest earthquake the fault is "competent" of generating is tacit to happen at the place on the fault adjoining to the site.

Seismic hazard mapping are done by two different approaches: a) Deterministic and b) probabilistic. In deterministic seismic hazard analysis (DSHA), earthquake scenarios are evaluated separately. For each sources (single faults or fault zones), a scenario earthquake is defined by magnitude, distance between source and area, style of faulting and in some cases rupture direction. The ground motion for the scenario earthquake is usually estimated by using attenuation relationship, but is sometimes estimated using seismological simulations of the ground motion. DSHA is based on geology and is attenuated to physical reality in nature. A typical DSHA can be described in four-step process:

1. Identification and characterization of all earthquake sources capable of producing significant ground motion at the site. Source characterization includes definition of each source's geometry and earthquake potential.

2. Selection of a source-to-site distance parameter for each source zone. In most DSHAs, the shortest distance between the source zone or point and the site of interest is selected. The distance may be expressed as an epicentral distance or hypocentral distance, depending on the measure of distance of the predictive relationships used in the following step.

3. Selection of the controlling earthquake (i.e, the earthquake that is expected to produce the strongest level of shaking), generally expressed in terms of some ground motion parameter. The selection is made by comparing the level of shaking produced by earthquakes (identified in step 1) assumed to occur at the distances identified in step 2. The controlling earthquake is described in terms of its size (usually expressed as magnitude) and distance from the site.

4. The characteristics of hazard are usually described by one or more ground motion parameters obtained from predictive relationships. Peak ground acceleration, peak velocity and response spectrum ordinates are commonly used to characterize the seismic hazard.

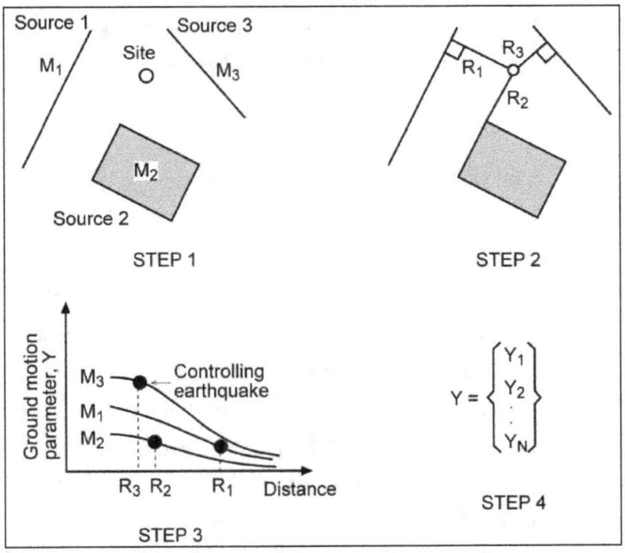

Four steps of deterministic seismic hazard analysis.

The probabilistic seismic hazard assessment (PSHA) involves the production of contour maps that represent the levels of ground and structural shaking expected to be experienced over a particular return period (in years). Generally, the shaking is represented in terms of peak ground acceleration (PGA) that is generated by the effects of the rupture of a geological fault and the propagation and attenuation of seismic waves reaching the bedrock. The shaking can be also represented by the structural acceleration (SA), and it depends on the period of vibration and damping of the building, and it is related to the characteristics of the ground motion itself; the SA is also referred in the literature as a spectral ordinate, and the damping is set as 5% of the critical (e.g., reinforced concrete structures) in PSHA computations. The PGA is associated with a zero structural period or with an infinite rigid structure. Then the severity of the shaking computed in a PSHA is primarily calculated on rock site conditions (NEHRP B soil class with a $V_s = 760$ m/s) and assuming a flat topography. The amplification of ground motion due to the presence of sediments above the bedrock is taken into consideration as a site effect in a second phase of evaluation.

The conjunction of spectral accelerations is represented in a response spectrum that is indeed the seismic coefficient *CS* that is always prescribed in the codes' regulations as a fraction of the gravity *g* (9.8 m/s²), i.e., 1.19 g = 11.66 m/s² at 0.2 s period. A common misunderstanding is to assume that the seismic coefficient is related only to the ground motion or PGA; such confusion comes from the fact that the shear base *V* of the building is related with the seismic coefficient employing the following equation:

$$V = C_S W,$$

where *W* is the total weight of the building. What equation $V = C_S W$ really reflects is the reaction at the ground level in the foundation of the building due to the all inertial forces *F* that result at each mass story subjected to its correspondent spectral acceleration SA during an earthquake. When considering a one degree of freedom system or a one-story building, the inertia force *F* can be calculated by the second Newton law as,

$$F = ma,$$

where m is the mass and a is the spectral acceleration. For the one-story building case $V = F$, $a = CS$, and $m = W/g$. Engineers use the weight instead of mass for structural calculations, so the seismic loads are directly calculated vanishing the gravity g constant when using $F = ma$: $F = (W/g)$*1.19 g. In the case of having a several degrees of freedom system or a multistory building, equation $V = C_S W$ also represents the reaction at the base of the building due to all inertia forces above; its application is still valid since spectral ordinates are always computed considering a single degree of freedom system that represents the seismic demand of one mode of vibration in the structure; in other words, the shear base in equation $V = C_S W$ retains the total inertia forces that are distributed later among all the stories along the building height when considering multi-degree of freedom systems. When considering the first overtones or harmonics in the calculations, the seismic demand of each mode of vibration can be combined employing classical methods of structural dynamics based on the same response spectrum derived by seismic hazard computations.

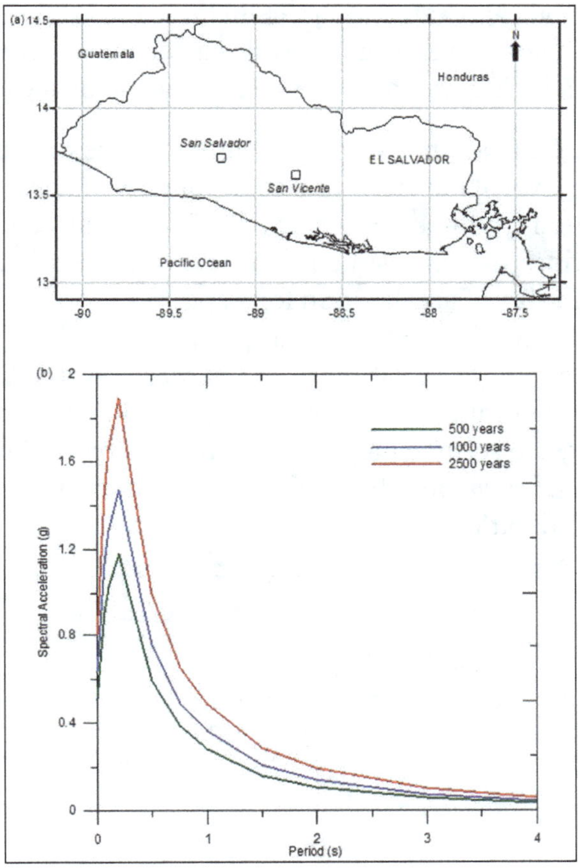

(a) Geographical location of El Salvador. (b) Elastic design response spectrum for 5% of critical damping at San Vicente City setting several years of return period.

Principles of Probabilistic Seismic Hazard Assessment (PSHA)

In PSHA calculations, the return period *RP* is associated with the expected level of shaking *CS* at one place and not with the recurrence interval *RI* of a single earthquake related with a specific magnitude in a seismogenic source. Despite the concepts of the return period *RP* and the recurrence interval *RI* are different they complement each other to produce the seismic loads that depend of the selected lifetime *L* (in years) of the building. The seismic coefficient *CS* in the modern code provisions is usually calculated setting 500- or 2500-year return period which is associated with the probability of exceedance: this is the probability that the building experiment at least one time—might be two or more—a specific level of shaking in its lifetime *L*. Employing a Poisson distribution, the probability of exceedance *q* for a specific motion level can be expressed as,

$$q = 1 - e^{-LE(z)},$$

where *E(z)* is the mean annual rate (number of earthquakes per year) for a specific level of motion *z*. Setting *q* to 2 and 10% and *L* = 50 years, *E(z)* yields 0.0004 and 0.002 earthquakes per year after evaluating equation $q = 1 - e^{-LE(z)}$. The return period is the reciprocal of *E(z)*; *RP* = 1/E(z) yields 2500 and 500 years in these cases. The lifetime of 50 years is presumably set as the average time span for which a person can own a building property. The multiplication of *L* by *E(z)* vanishes the unit of years allowing probability results *q*.

Then, the main objective of seismic hazard computations is to obtain the mean annual rate *E*(z) for a specific level of motion z, and consequently the probability of exceedance is calculated in equation $q = 1 - e^{-LE(z)}$ setting a lifetime usually set as *L* = 50 years. For a specific period of vibration, a seismic hazard curve is calculated at several motion levels, and the design value of acceleration for a specific period of vibration is obtained at 2 or 10% of probability of exceedance via interpolation of the hazard curve. When considering another period of vibration in the hazard computations, it is possible to construct a response spectrum related to the engineering project at the return period of interest. Finally, the seismic hazard maps are constructed using a grid of points smoothing the acceleration values obtained at each point that constitutes the grid over the region under consideration.

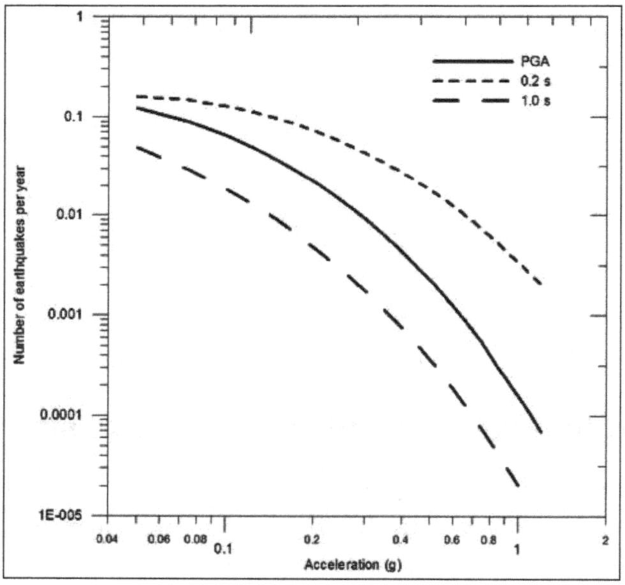

Seismic hazard curves depicting the number of earthquake per year vs. the peak ground acceleration and spectral accelerations for 0.2 and 1.0 s structural period of vibration.

Rigid-zone Method

The rigid-zone method is the standard Cornell-McGuire approach based on a Poissonian probability distribution for which the occurrence of an event is independent of others. The first step in PSHA is to quantify the level of seismicity in the region of study answering three questions: (i) Where do earthquakes happen? (ii) What magnitude do they have? and (iii) How do they frequently occur? Several *ith* seismogenic sources must be declared as polygons that enclose a group of earthquakes in particular geographical location that have similar geological characteristics, namely, the tectonic environment, the focal mechanism, the focal depth, and a maximum magnitude. As a result, that area constitutes a rigid zone. It is called the rigid zone because it is assumed that the level of seismicity is constant all over the area inside the polygon and cannot change with time and dimensions. Since the hazard analysis is based on a Poisson probability distribution, the compiled earthquake catalog must be declustered, that is, the foreshocks and the aftershocks must be removed from the list of events, and only the main shocks must remain in the catalog; this is usually accomplished employing decluster algorithms. To quantify the seismicity in a specific seismogenic source, a Gutenberg-Richter (G-R) recurrence relationship is defined as,

$$\log N_i = A - BM,$$

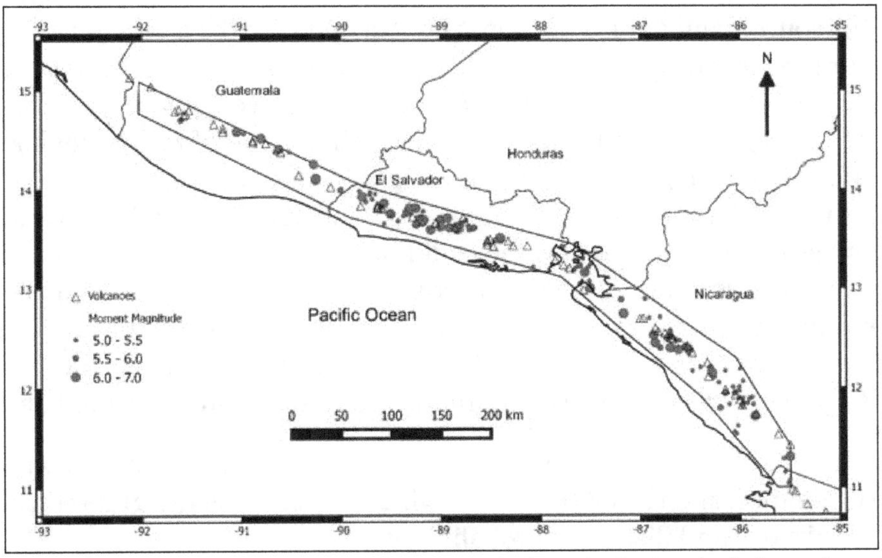

Epicenters (solid circles) and geometrical delimitation (polygon) of volcanic chain earthquakes in El Salvador and neighboring countries. The active quaternary volcanoes are depicted with white triangles. In this Figure the epicenters of main events, foreshocks, and aftershocks are included.

where *Ni* represents the cumulative number of earthquakes per year above magnitude *M*, and *A* and *B* set two constants derived from regression analysis; such constants are independent of the period of vibration of buildings. The number of earthquakes *Ni* within different magnitude ranges are usually computed employing the bounded G-R relationship. The constant *-B* is the measure of the relative abundance of large to small shocks, and it includes the negative sign indicating the negative slope that must result from the regression analysis. Equation $\log N_i = A - BM$ must be derived after a completeness analysis of the earthquake catalog; what it means is that the cumulative number of earthquakes must be divided by the number of years for which a particular magnitude bin is completed. It is worth mentioning that the reciprocal value of *Ni* yields a recurrence interval *RI* and does not yield the return period for a certain magnitude *M*. For example, if $A = 5.2$ and $B = 1$ for a *M* 4.0 and *M* 7.0, N_i yield 16.0 and 0.016 earthquakes/year, and the recurrence interval are ¾ and 63 years, respectively. Then the return period is associated with the level of motion at a particular site, and the recurrence interval is related with the size of an earthquake.

Although this information is valuable in terms of occurrence characteristics of the earthquakes sources, such recurrence interval is still not associated with a specific ground or structural motion needed to design a building. In the first instance, a ground motion prediction equation (GMPE) must be developed in order to relate the expected shaking at one place of interest depending on the magnitude and distance from the source at rock site conditions. Such GMPE must be compatible with the same tectonic environment of the seismogenic area *ith* to develop equation $\log N_i = A - BM$. A general form of a GMPE is,

$$\log(y) = a + bM + cR + d\log R \pm \sigma_{\log(y)},$$

where y is the PGA or spectral acceleration SA in g, a and b are regression constants related to the source, and c and d are constants related with the an elastic and geometrical attenuation respectively (path term). M is the moment magnitude, R is the distance from the project site to any point of the seismogenic source i, and $\sigma_{\log(y)}$ is the standard deviation of the regression in terms of the logarithm value of y. Employing equation $\log(y) = a + bM + cR + d\log R \pm \sigma_{\log(y)}$ and setting a = 1.68, b = 0.30, c = −0.01, and d = −1.0, the estimated PGA for an earthquake of M = 7.0 at a distance R = 10 km gives 479 cm/s² = 0.49 g. Despite that such value could be used for the design of a short period structure, it is not associated with any recurrence interval *RI* of this earthquake size and with any lifetime L and return period *RP*. Moreover, there is no information of other magnitude and distance pairs that can affect the project site; besides, the standard deviation σ is not being taken into account in the ground motion estimation at this stage.

The mean annual rate *E(z)* for a specific motion level "z" can be obtained combining the seismicity evaluation and the GMPEs employing the following integrals considering all possible magnitudes and distances probabilities:

$$E(z) = \sum_{i=1}^{Ns} v_i \int_{R=0}^{\alpha} \int_{M_{min}}^{M_{max}} (M) f_i(R) P(Z > z \mid M, R)\, dR\, dM,$$

where Ns = number of seismogenic sources; v_i = mean rate of occurrence of earthquakes above the minimum magnitude for the "ith" source; fi(M) = probability density distribution of magnitude within the "ith" source, which is obtained using the G-R relationship; $f_i(R)$ is the probability density distribution of epicentral distance "R" between various locations within source "ith" and the site where hazard is estimated; and P(Z > z |M,R) = probability that a given earthquake of magnitude "M" and source-site distance "R" will exceed motion level "z", which is obtained employing the selected GMPE and the standard deviation σ.

Events with a lower magnitude Mmin (i.e., 5.0) might not cause damage to structures, so seismic hazard computation is performed establishing a lower bound limit in all cases. The mean rate of occurrence v_i is evaluated using equation $\log N_i = A - BM$ at M_{min} and represents all earthquakes per year that could happen in the seismogenic source above the lower bound limit nearby the project location, including the maximum credible magnitude M_{max} that can produce each seismogenic source ith; M_{max} is considered as the upper-bound physical limit in hazard computations. The units of E(z) are in accordance with the units of v_i and must be multiplied by the results of the integrals that contain the unitless quantities represented by the probabilities of occurrence of magnitudes, distances, and the level of shaking z produced by all possible magnitude-distance pairs that affect the site of interest. The multiplication of $f_i(M)$, $f_i(R)$, and $P(Z > z \mid M,R)$ obeys to the fact that the occurrences of these probabilities are physically and statistically independent.

The probability distribution of magnitude M or the cumulative probability function CDF(M) can be expressed as,

$$CDF_i(M) = \frac{N_{M_{min}} - N_M}{N_{M_{min}}}.$$

The *CDFi(M)* is the quotient between the numbers of earthquakes of a magnitude *M* that exceeds the number of earthquakes of minimum magnitude *Mmin* after evaluation in Equation $\log N_i = A - BM$. However, the derivation of equation $CDF_i(M) = \frac{N_{Mmin} - N_M}{N_{Mmin}}$ with respect to the magnitude *M* is used in seismic hazard calculations, yielding the probability density distribution after Neperian logarithm transformation of the *B* value in the G-R relationship:

$$f_i(M) = \beta e^{-\beta(M - M_{min})},$$

where β = (ln10)B and e is the Neperian number. The probability density function is suitable for PSHA since it represents the chances of success of the occurrence of particular size M of an earthquake. In computational practice, Equation $f_i(M) = \beta e^{-\beta(M - M_{min})}$ is discretized over the same bin magnitude classes (e.g., M_1 5.5 to M_2 6.0) used to develop the G-R recurrence relationship (equation $\log N_i = A - BM$); then the probability P(M) of a magnitude to fall in that bin can be written as,

$$P(M) = \int_{M_1}^{M_2} f_M dM$$

where the value of *M* is taken as the average of the bin magnitudes *M1* and *M2*.

When performing PSHA calculations, the polygon that represents the rigid zone is usually discretized in squares of 1 km side which represent all possible earthquake sources. The probability $f_i(R)$ that an earthquake source could happen at a specific distance R from the project site is the quotient between R and all the possible distances R_j:

$$P(R) = f_i(R) = \frac{R_j}{\sum_{j=1}^{n} R_j}$$

where n is the number of elements used to discretized the seismogenic source. All the earthquake sources can have magnitudes between the lower- and upper-bound limits with related probabilities of occurrence explained above. So, $f_i(M)$ and $f_i(R)$ are independent of the level of motion being evaluated; $f_i(R)$ is dependent on the location of the project relative to the seismogenic source ith into consideration. The probability $P(Z > z \mid M,R)$ that a given earthquake of magnitude M and source-site distance R will exceed motion level z is evaluated using a standard normal variable G:

$$G = \frac{\log(z) - \log(y)}{\sigma_{\log(y)}}.$$

To estimate $\log(y)$ in equation $\log(y) = a + bM + cR + d\log R \pm \sigma_{\log(y)}$ and $G = \frac{\log(z) - \log(y)}{\sigma_{\log(y)}}$, the value of M to be used in equation $\log(y) = a + bM + cR + d\log R \pm \sigma_{\log(y)}$ is taken as the average of M_1 and M_2 in equation $P(M) = \int_{M_1}^{M_2} f_M dM$. $P(Z > z \mid M,R)$ is then calculated as,

$$P(Z > z \mid M,R) = 1 - \phi_{(G)},$$

where $\phi(G)$ is the probability density function derived from a log-normal Gaussian distribution evaluated for the *G* value obtained in equation $G = \frac{\log(z) - \log(y)}{\sigma_{\log(y)}}$. For computational purposes, the exact integrals in equation $E(z) = \sum_{i=1}^{Ns} v_i \int_{R=0}^{\alpha} \int_{M_{min}}^{M_{max}} (M) f_i(R)$ $P(Z > z \mid M,R)\, dR\, dM$ are substituted by discrete sums yielding:

$$E(z) = \sum_{i=1}^{Ns} v_i \sum_{R_{min}}^{R_{max}} \sum_{M_{min}}^{M_{max}} P(M) P(R) P(Z > z \mid M,R).$$

The probabilities in equation $E(z) = \sum_{i=1}^{Ns} v_i \sum_{R_{min}}^{R_{max}} \sum_{M_{min}}^{M_{max}} P(M) P(R) P(Z > z \mid M,R)$ are calculated after selecting a segmentation of magnitude and source-site distance pairs. Usually such segmentation is done for every 0.25–0.5 magnitude units of magnitude and 1–10 km distance. *Rmin* and *Rmax* constitute the minimum and maximum source-site distances.

The Free-zone Methods

Woo proposed an alternative approach to compute the PSHA in order to remove the uncertainties inherent in the definition of the rigid seismogenic sources above. The arbitrariness in delineating zone boundaries may be artificially expanded to dilute activity around a site or on the other hand artificially contracted to concentrate activity around a site. Besides, the assumption that the activity rate is constant in a seismogenic source is rarely tested statistically. In this method, the seismic hazard is calculated directly based on the characteristics of the earthquake catalog rather than defining rigid zones considering a spatial uniformity seismicity; then earthquake sources are not defined according to a geological and tectonic criteria. The Woo method is called the kernel estimation method since it implements spatial smoothing directly to earthquake epicenters contained in the catalog depicting the spatial non-uniformity of seismicity in the region of study. Then the earthquake catalog must contain the main events, the foreshocks, and aftershocks when performing hazard calculations.

The first step in the method consists in defining a grid of nodes that constitute the point earthquake sources around the project site. Each point of the grid would have an activity rate that is magnitude-distant dependent based on fractal characterization. A multivariate probability density function *K* is expressed in terms of magnitude and distance yielding:

$$K(M,x) = \frac{n-1}{\pi H^2}\left[1+\left(\frac{R}{H}\right)^2\right]^n,$$

where *n* is the exponent of the power law (usually between 1.5 and 2.0), *R* is the source-site distance, and *H* is the bandwidth for normalizing distances that is a function of magnitude *M* representing the average distance between epicenters of the same magnitude bin:

$$H = ce^{dM},$$

where c and *d* are constants to be derived by a regression analysis. By summing over all cataloged events with historical epicenters *xi*, the cumulative activity rate density is computed at a general point *x* in the grid for each magnitude bin class from the minimum magnitude to the highest value, as follows:

$$\lambda(M,x) = \sum_{i=1}^{n_o} K(M, x - x_i) / T(x_i).$$

T(xi) is the effective observation time period which is equivalent to the completeness period, and *no* is the number of earthquakes in the catalog; the contribution of each event is inversely weighted by its completeness period. Once *λ(M,x)* is calculated for each node in the mesh, a GMPE is employed, and the seismic hazard is calculated by summing over each node source for an specific motion *z*. The probability *P(Z > z |M,R)* will be evaluated using equation $G = \frac{\log(z) - \log(y)}{\sigma_{\log(y)}}$ and $P(Z > z | M,R) = 1 - \phi_{(G)}$ and multiplied by *λ(M,x)* to calculate *E(z)* and the correspondent probability of exceedance *q* in equation $q = 1 - e^{-LE(z)}$ setting a specific lifetime *L*. The observational uncertainties in event magnitude and epicentral location can be accounted for in this method employing Gaussian error distributions. The basic principle underlying the Woo method is that the epicenter of each past shallow event is smoothed geographically to generate a spatial probability distribution for event recurrence. As a conclusion, the method intends to represent both the fractal clustering of moderate magnitude epicenters and the haphazard migration of major events.

Frankel also proposed a free-zone method used for the US National Hazard Maps developed by the US Geological Survey (USGS). In this method a region is partitioned into cells by a grid of 0.1° (11 km on a side). Within the *i*th grid cell, a count is made of the number of events with a magnitude above a reference value *Mref*. This cumulative count of events is converted to incremental values from *Mref* + *ΔM* or for a specific magnitude bin. The grid of *ni* values is smoothed spatially in a simple Gaussian way with correlation distance *cd*. The smoothed value of *ni* is obtained as follows:

$$\frac{\sum n\, e^{-\Delta^2 ij/cd^2}}{\sum {}^{-\Delta^2 ij/cd^2}},$$

where *Δij* is the distance between the ith and jth cells. For a specific site, the values of n_i are binned by their distance from the site, and a value of N_r is calculated as the total value of n_i for cells within a certain distance increment of the site.

The mean annual rate E(z) for a specific motion level "z" at the project site is determined as follows:

$$E(z)=\sum_r \sum_m 10^{\left[\log\left(\frac{N_r}{T}\right)+B\left(M_m-M_{ref}\right)\right]} P(Z>z|M,R)$$

where *r* and *m* are the counters for the distance and the magnitude bins, respectively.

It is worth mentioning that the term $10^{\left[\log\left(\frac{N_r}{T}\right)+B\left(M_m-M_{ref}\right)\right]}$ is the value of 10^A in equation $\log N_i = A - BM$ for the completeness period *T*; $P(Z>z|M,R)$ is calculated as in the previous methods. Since a Poisson probability distribution is used in this method, the earthquake catalog must be declustered removing the foreshocks and aftershocks retaining only the main events for PSHA calculations.

The Characteristic Earthquake Model

The characteristic earthquake model is employed when detailed fault information is collected by a geologist in the field and where there is a clear evidence of the slip via geodetic surveys (GPS) and/or paleoseismological investigations. Instead of declaring an area as a rigid seismogenic source or employing fractal geometry concepts in free-zone methods, a fault is defined employing the following input data:

- The slip rate (mm/year).
- A *B* value (usually set as 1.0).
- Precise geographical coordinates of a series of points that define the fault trace.
- The dip, the sense of slip, and depth of the fault.
- The minimum magnitude *Mmin* (e.g., 6.0) to be considered in the analysis.

Two alternate characteristic models are applied actually in a PSHA:

- Considering only the maximum magnitude *Mmax* and its associated activity rate or number of earthquakes per year.
- Classical Gutenberg-Richter (G-R) relationships.

For both models the first step in a PSHA is to calculate the number of earthquake per year *E(z)* for the *Mmax* in model (i) and for the *Mmin* in the model (ii).

The idea of using two alternate models is that it is unknown whether future earthquakes will be large fault-filling ruptures of smaller ruptures which only occupy a relatively small portion of the fault. Generally a weight of 0.5 is assigned to each model if the mean recurrence time for *Mmax* in the first model is less than the return period of motion.

Probabilistic Seismic Hazard Application for the Volcanic Chain Earthquakes in El Salvador

Here, free-zone method, the rigid-zone methods, and the characteristic models are applied for the volcanic chain earthquakes in El Salvador. For the Cornell-McGuire method, an own program was developed and used the KERFRACT program for the Woo method. The programs developed by the USGS were employed for the Frankel and characteristic models. The seismic activity of this source is concentrated in upper part of the crust with focal depth less than 25 km and within a continuous belt of 20–30 km width along the active volcanoes in Central America. The events can appear in clusters. Faults with right and left lateral focal mechanisms are located parallel and perpendicular to the volcanic axis, respectively.

Seismic hazard maps for PGA were developed employing a 0.1 × 0.1°° grid assuming a flat topography setting 500-year return period of ground motion in terms of PGA. The homogenized earthquake catalog compiled by Salazar et al. was used for the moment magnitudes above 5.0 (*Mmin*) and the GMPE developed by Sadigh et al. for shallow crustal earthquakes and a truncation value of 3σ in the hazard computations. We updated the work of Bommer et al. which also presents PSHA maps for El Salvador in terms of normalizing PGA to the maximum computed value; here the absolute values of PGA are presented based on new seismological information applying the methods explained above.

A *Mmin* of 5.0 and *Mmax* of 6.9 were set when using all methods with a constant seismogenic depth of 5 km. The G-R relationship for the rigid-zone method is presented in figure after the completeness analysis. A tri-linear trend that reflects interesting features of the seismicity along the volcanoes is observed. The first part (B_1) suggests a large number of small earthquakes related to big shocks, the middle part yields a very low B_2 value reflecting the effects of clusters or the abundance or moderate size events, and the third part (B_3) reflects the small number of destructive events related to small shocks. The tri-linear trend of the G-R relation yields the probability density function when applying equation $CDF_i(M) = \frac{N_{M_{min}} - N_M}{N_{M_{min}}}$ and $f_i(M) = \beta e^{-\beta(M-M_{min})}$. The sum of all magnitude bin probabilities of the tri-linear trend yields 76, 9, and 15% for the first (B_1), second (B_2), and third (B_3) segments, respectively, which sum must be 100%. Although, the probability of occurrence of an earthquake of a specific magnitude M 5.6 and M 6.3 is the same of about 3%, presumably, the contribution to the seismic hazard from both events in terms of magnitude probability is the same despite the size differences of these events . Generally, the greater the magnitude, the lower the probability. The probability of occurrence for magnitudes between 5.7 and 6.1 yield similar (2%) due to the low B2 value in the G-R relationship.

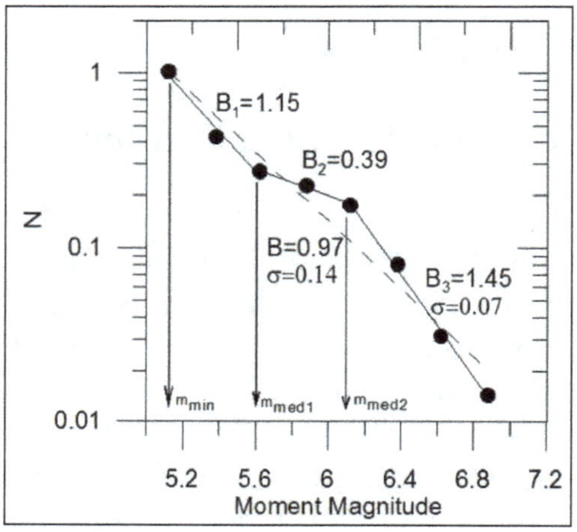

Gutenberg-Richter (G-R) relationships for the volcanic chain earthquakes. Solid line and solid circles: Tri-linear trend employed in the rigid-zone Cornell-McGuire method. Broken line: Linear regression employed in the free-zone Frankel method with B value of 0.97.

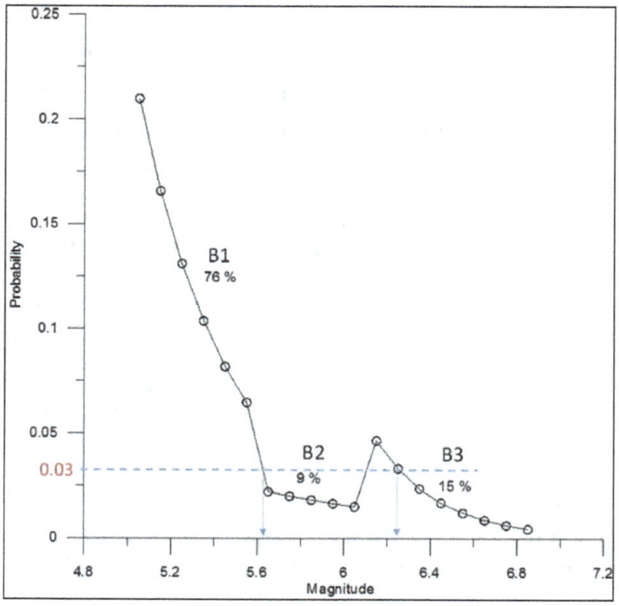

Magnitude probability density function for the volcanic chain earthquakes based on the tri-linear G-R relationships. The percentages are the sum of all magnitude bin probabilities for each segment of the tri-linear trend with a total of 100%.

Carr et al. suggest that the volcanoes in Central America cluster into centers, whose spacing is random but averages about 27 km. What it means is that earthquake epicenters and volcano centers have the same spatial distribution among them. Regarding with the Woo free-zone method, the kernel only depends on the epicentral distance rather than magnitude, so the kernel function must be unique for the volcanic chain events and independent of the size of the earthquakes; then c = 27 and d must be zero yielding H = 27 km in equation $H = ce^{dM}$.

Table: Average distances for magnitude bins for volcanic chain earthquakes.

Magnitude range	Number of earthquakes	Distance (km)
$5.00 \le M < 5.25$	68	15.84
$5.25 \le M < 5.50$	13	35.19
$5.50 \le M < 5.75$	20	22.33
$5.75 \le M < 6.00$	14	24.79
$6.00 \le M \le 6.25$	22	19.44
$6.25 \le M \le 6.50$	11	22.64
$6.50 \le M \le 6.75$	5	47.98
$6.75 \le M \le 7.00$	2	27.75
Total	155	Average H = 27

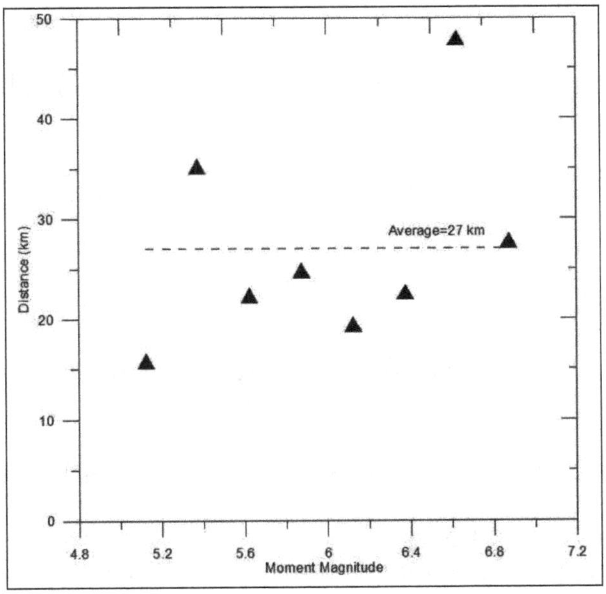

Bandwidth for normalizing epicentral distances in El Salvador volcanic chain and neighboring countries where H = 27 km in equation $K(M,x) = \frac{n-1}{\pi H^2}\left[1+\left(\frac{R}{H}\right)^2\right]^n$ and

$H = ce^{dM}$. Carr et al. suggest the volcanoes in Central America clusters into centers, whose spacing is random but averages about 27 km. It is noted that the all average distance for all earthquake epicenters also yields the same 27 km as the volcano centers.

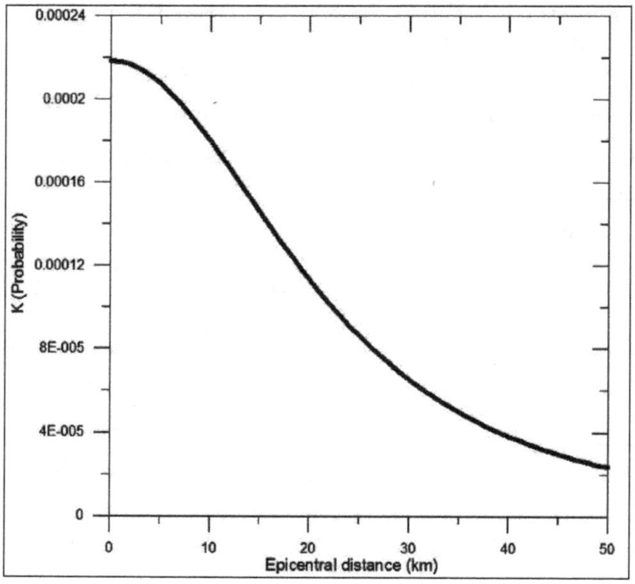

Kernel function for upper-crustal earthquakes in the volcanic chain. The kernel probability function K is unique for H = 27 m in equation $K(M,x) = \frac{n-1}{\pi H^2}\left[1+\left(\frac{R}{H}\right)^2\right]$ and $H = ce^{dM}$.

We model the characteristic earthquake for the San Vicente Fault employing the following parameters:

- The slip rate: 4.1 mm/year based on paleoseismological investigations.
- A *B* value equal to 1.0.
- The geographical coordinates that define the fault trace are taken from.
- The dip = 90°, the sense of slip as a pure right lateral and depth of the fault = 15 km.
- The minimum magnitude *Mmin* as 6.0 to be considered in the analysis.

The seismic hazard maps employing the four methods explained above. As a first observation, it is noted that the rigid-zone method yields a uniform level of ground motion along the volcanoes and that the free-zone methods can capture the nonuniformity of hazard spatial distribution especially at the western part of the country. However, at San Salvador City, the PGA yields about 0.45–0.5 g for all methods. At San Vicente City when applying the rigid zone, the Woo free zone, and the characteristic model, the PGA yields 0.45 g; however, a lower value of about 0.35–0.40 g yields for the Frankel free-zone method. This difference is attributed to the fact that the Woo method employs the unclustered catalog and the Frankel method employs the declustered one that it only contains the main events; it is noticed also that for the Frankel method, we employed a unique B value of 0.97 for a G-R relationship; however, such hazard estimates are not significantly different from each other. When applying the characteristic model employing the paleoseismological slip rate at San Vicente City of 4.1 mm/year, a near value of 0.45 g is obtained as well. The coincidence of the PGA values revealed the quality of our earthquake catalog in the volcanic chain. The actual state of the art suggests to treat the epistemic uncertainties in the PSHA through a logic-tree framework considering all the methods and compute the covariance or error associated with the result of each branch in the logic tree. The PSHA could be performed for spectral ordinates as well; figure depicts an elastic response spectrum for San Vicente City employing the Woo free-zone method for 13 periods of vibration between 0.0 and 4.0 s setting 5% of critical damping.

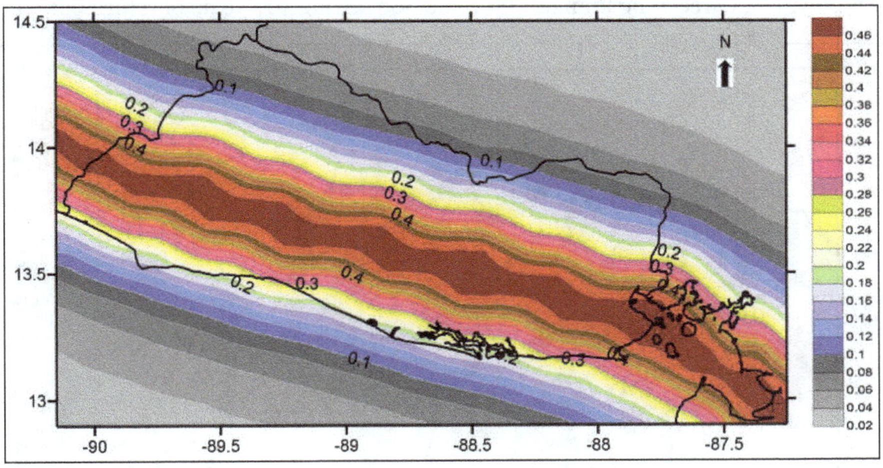

Seismic hazard map for the peak ground acceleration (PGA) as a fraction of gravity "g" employing the rigid-zone Cornell-McGuire method setting 500 years of return period.

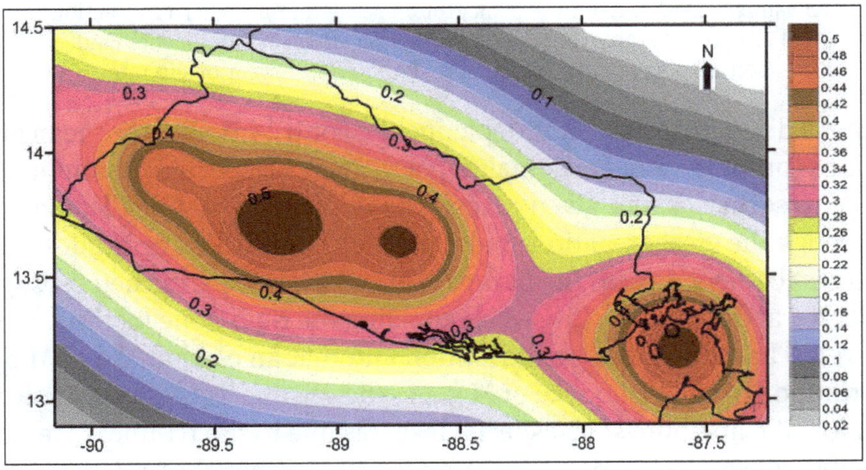

Seismic hazard map for the peak ground acceleration (PGA) as a fraction of gravity "g" employing the free-zone Woo method setting 500 years of return period.

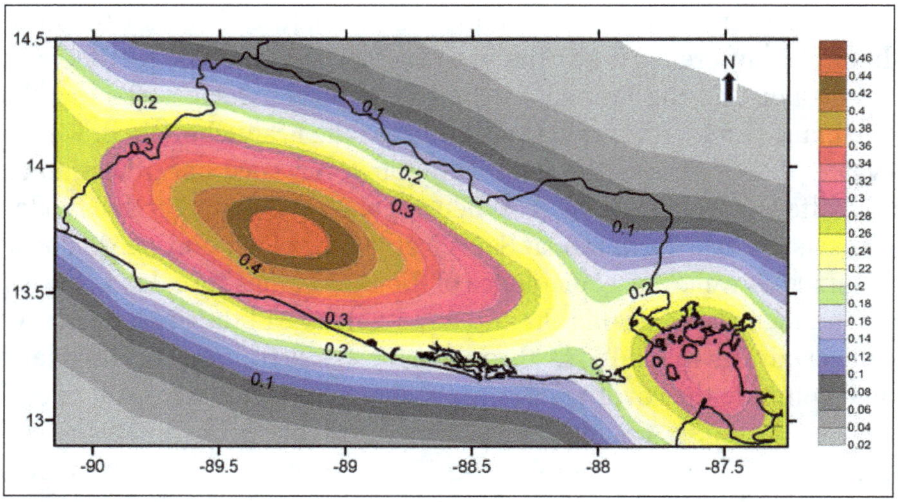

Seismic hazard map for the peak ground acceleration (PGA) as a fraction of gravity "g" employing the free-zone Frankel USGS method setting 500 years of return period.

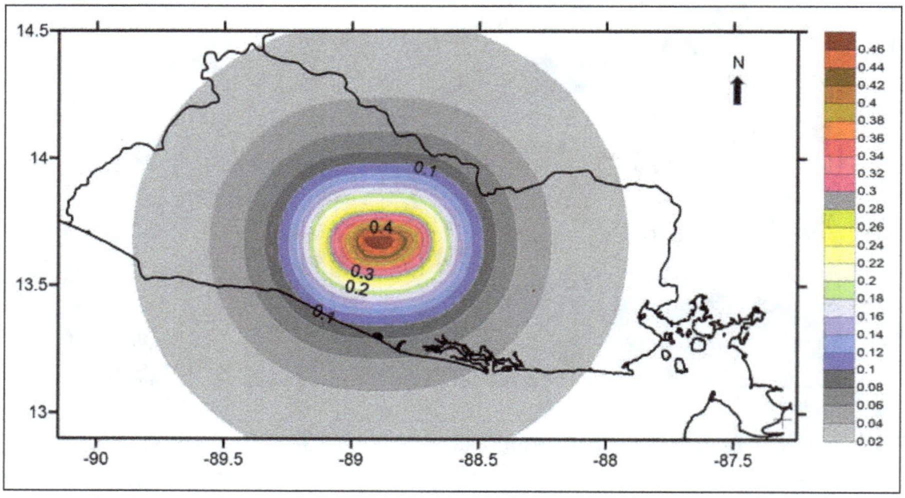

Seismic hazard map for the peak ground acceleration (PGA) as a fraction of gravity "g" employing the characteristic model at San Vicente Fault setting 500 years of return period and 4.1 mm/year of slip rate.

Figure shows a disaggregation chart depicting the contribution to the mean annual rate of exceedance (or number of earthquakes per year) for a specific ground motion value at a site due to scenarios of a given magnitude and source-to-site distance. Then it is possible through the disaggregation analysis to determine the magnitude-distance pair with the largest contribution to the hazard. This analysis is performed setting a PGA of 0.46 g observed at San Vicente City when applying the rigid-zone method. Small earthquakes of M 5.0 have the largest contribution and that an earthquake of M 6.2 has the same contribution of an earthquake of M 5.4 to the seismic hazard in the volcanic chain of El Salvador. We attribute such especial features due to the tri-linear trend observed in the seismicity evaluation represented by the G-R recurrence relationships and its correspondent magnitude probability density function presented in figure.

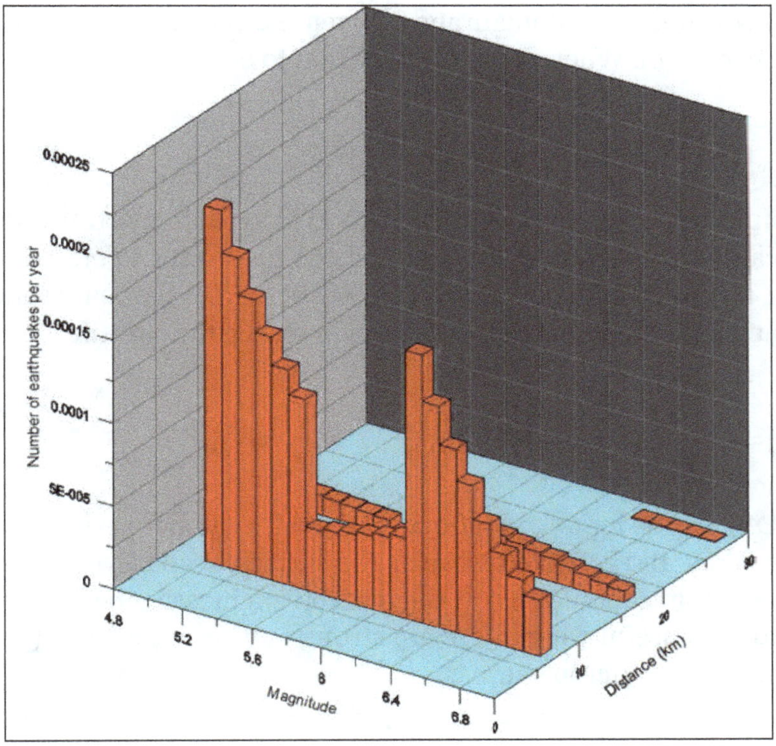

Example of magnitude-distance pair disaggregation charts for the peak ground acceleration (PGA) of 0.46 g at 500-year return period employing the rigid-zone Cornell-McGuire method at San Vicente City. An earthquake of M 6.2 has the same contribution to seismic hazard of an earthquake of M 5.4 in the volcanic chain of El Salvador.

Site Effects Related with the Interbedding of Sediments and Lava Flows

Different studies carried out after the occurrence of several destructive seismic events have confirmed that the presence of sedimentary deposits exerts a particularly strong influence on the distribution of damages. These studies have also corroborated that the spatial distribution of damages correlates on the magnitude of the amplifications of the arriving seismic waves caused by the upper sedimentary layers. Consequently, we can regard the distribution of the ground-shaking intensity as a phenomenon closely related to the filtering effects of the soil profile above the bedrock level .

It is convenient to introduce the most general and widely accepted formulation of the ground motion phenomenon, departing from assuming that in the frequency domain, the S-wave can be expressed in terms of Fourier amplitude spectra *Aie(f)* recorded at a site *i* due to an earthquake *e* as,

$$A_{ie}(f) = S_e(f)P_{ie}(f)G_i(f),$$

where f denotes the frequency domain and $S_e(f)$ represents the source, $P_{ie}(f)$ the path, and $G_i(f)$ the site effect term. When applying logarithm to equation $A_{ie}(f) = S_e(f)P_{ie}(f)G_i(f)$ yields,

$$\mathbf{logA_{ie}(f) = logS_e(f) + logP_{ie}(f) + logG_i(f)}$$

where the amplitude term *Aie(f)* is related to PGA or and spectral ordinate SA, the source term can be converted to *a* + *bM* and the path term to *cR* + *dlogR* in equation $\log(y) = a + bM + cR + d\log R \pm \sigma_{\log(y)}$. The convolution of the source and path effects yields the input bedrock motion *Bn*; in the first instance the site effect term can be defined as,

$$\mathbf{G_i(f) = \frac{B_m + A_m}{B_n} = \frac{2B_m}{B_n}}$$

where *Bn* and *An* are the input and refracted bedrock motion, respectively, and *Bm* and *Am* are the incident and refracted wave at the surface, respectively. The free-surface effects yield an amplification of 2.0 at the outcrop engineering bedrock since $B_n = A_n$. Such formulation is useful when there are strong motion recordings (accelerograms) for both sediment and outcrop sites.

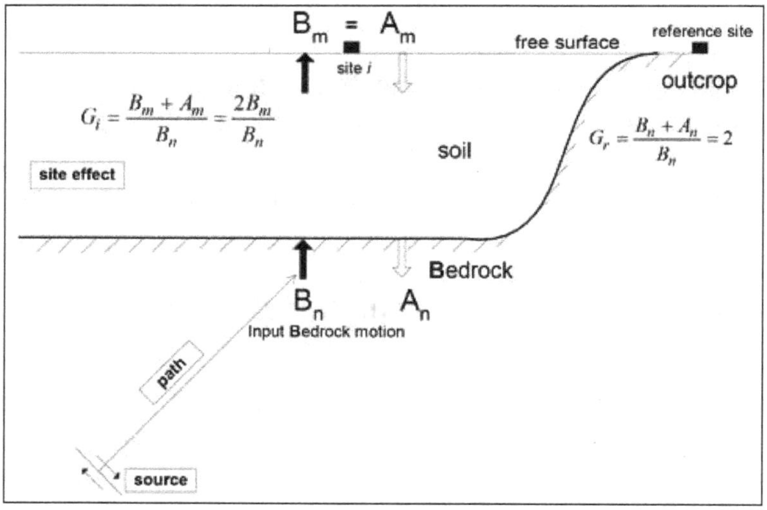

Illustration of theoretical transfer functions at soil sites and outcrop site taking into account the free-surface amplification. Bn and An: Input and refracted bedrock motion, respectively. Bm and Am: Incident and refracted wave at the surface, respectively.

The estimation of the magnitude of site effects applying theoretical methods requires knowledge of the physical properties of the layers composing the soil profile, namely, shear wave velocity, density, and thickness of the layers of the upper 30 meters. In most practical cases involving thick sedimentary deposits, knowledge of the thickness and physical properties of the upper 30 meters of the soil profile appears to provide a rather poor representation of the problem.

The amplification function employing 1-D SH wave propagation can be obtained when solving the following partial equation of wave motion:

$$\frac{\partial^2 u}{\partial t^2} = v^2 s \frac{\partial^2 u}{\partial v_v^2},$$

where *u* is the horizontal displacement, *vs* is the shear wave velocity, *t* is time, and *v* is the vertical direction.

A theoretical 1D SH amplification function is derived solving equation $\mathbf{G_i(f)} = \frac{\mathbf{B_m + A_m}}{\mathbf{B_n}} = \frac{\mathbf{2B_m}}{\mathbf{B_n}}$ and $\frac{\partial^2 u}{\partial t^2} = v^2 s \frac{\partial^2 u}{\partial v_v^2}$ for a soil profile volcanic environment at the capital of El Salvador, San Salvador. The work of Salazar and Seo is revisited analyzing a volcanic ash and rigid lava flows' interbedding patterns. A value of v_s = 225 m/s is used for the volcanic ashes in the upper part (*Tierra Blanca*) and a value of v_s = 500 m/s for the soils under laying the stiff lava flow (consolidated tuffs). In the case of the stiff lava flow, we used a v_s = 2100 m/s. The interbedding of soft and rigid materials is the consequence of different volcanic eruptions through historic times.

Interbedding of lava flow and volcanic ashes.

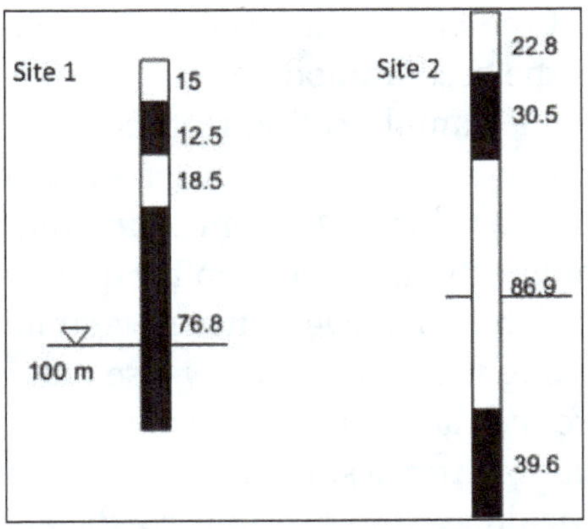

Profiles of lava flow and volcanic ashes interbedding. The value of thickness of the volcanic ash (white part) and the lava flow (black part) is written to the right of the profiles in meters.

1D SH wave transfer functions for interbedding of lava flow and volcanic ashes. (a) Site 1 and (b) Site 2 in figure. The second mode is predominant at 0.4 s rather than the first one for the Site 1 and that similar amplification factors of about 10–12 are obtained for the first and second modes for Site 2.

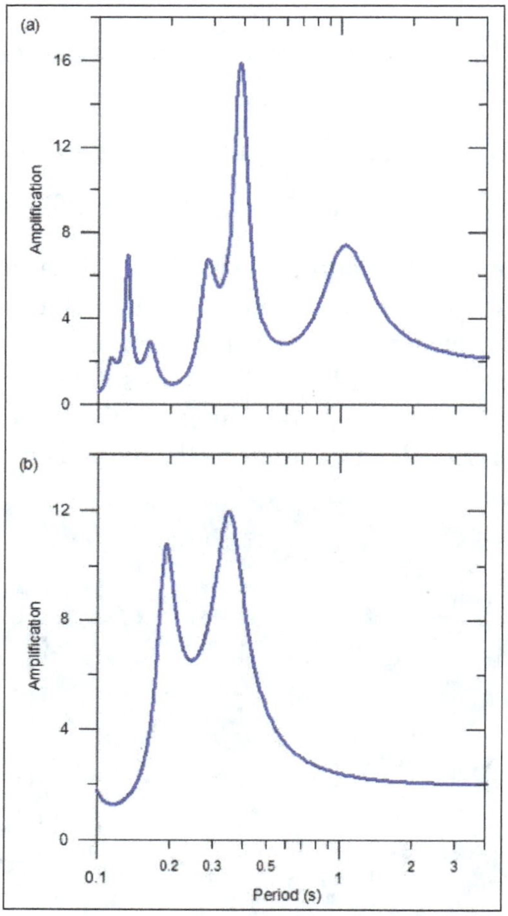

For the site 1, the highest amplification is observed at the second mode of vibration, and it does not take place at the first mode as can be usually expected, so in this case, the predominant period is the second mode of soil vibration at 0.4 s rather that the first one. For the site 2, similar amplification factors of about 10–12 are observed at the first mode and the subsequent overtone. For long period components above 2.0 s the amplification factors yield a value of 2.0 confirming that the earthquake motion at these periods yields no amplification. These amplification features might have serious implications in the distribution of damages since several modes can be highly excited during an earthquake affecting medium- and low-rise buildings in San Salvador City. Moreover, if the building foundations are being constructed on the first lava flow encountered as a recommendation from a standard penetration test (SPT), the site would be characterized as a rock yielding a lower seismic coefficient despite that volcanic ashes might underlay the lava several tens of meters.

ASSESSMENT OF SEISMIC HAZARD OF TERRITORY

The new complex method of seismic hazard assessment that resulted in creation of the probabilistic maps of seismic microzonation is presented in this chapter. To study seismicity and analyze seismic hazard of the territory the following databases are formed: macroseismic, seismologic databases and the database of possible seismic source zones (or potential seismic sources - PSS) as well. Using modern methods (over-regional method of IPE RAS - Russia) and computer programs (SEISRisk-3 – USA) in GIS technologies there were designed some probabilistic maps of seismic hazard for the Republic North Ossetia-Alania in intensity units (MSK-64) at a scale of 1:200 000 with exceedance probability being of 1%, 2%, 5%, 10% for a period of 50 years, which corresponds to recurrence period of 5000, 2500, 1000, 500 years. Moreover, first the probabilistic maps of seismic hazard were made in acceleration units for the territory of Russia. The map of 5% probability is likely to be used for the large scale building, i.e. the major type of constructions, whereas the map of 2% probability should be used for high responsibility construction only. The approach based on physical mechanisms of the source is supposed to design the synthesized accelerograms generated using real seismic records interpretation.

For each of the zoning subject the probabilistic map of the seismic microzonation with location of different calculated intensity (7, 8, 9, 9*) zones is developed (the zones, composed by clay soils of fluid consistency, which can be characterized by liquefaction at quite strong influences, are marked by the index 9*). The maps in acceleration units show the similar results.

The complex approach based on the latest achievements in engineering seismology, can significantly increase the adequacy or foundation for assessments and reduce the inaccuracy in earthquake engineering and construction.

Realization of investigations on mapping of seismic hazard such as detailed seismic zoning (DSZ) based on the most advanced field research methods and analysis of every subject of the Northern Caucasus separately on a scale of 1:200 000 gives the possibility to merge a bit unavailable, at first glance, schemes into geologically and geophysically quite reasonable map of DSZ for the Northern Caucasus with equal scale system of the source zones.

Assessment of Seismic Hazard: General and Detailed Seismic Zoning

The seismic hazard of some territory represents a possible potential or a level of expected hazard, caused by geological structure features, tectonic movements, geophysical fields, macroseismical catalog, engineer-geological and hydrogeological structure etc.

The adequate assessment of seismic hazard, at the same time is one of the important problems of engineer seismology. Unlike short-range and middle-range earthquake forecast, the involved assessment of seismic hazard, presented as seismic zoning maps, in fact is a long-ranged forecast of the earthquake strength and place.

One can mark out three types of analysis three consecutive stages of seismic zoning:

- General seismic zoning – GSZ or SZ, is realized in 1:5 000 000 or 1:2 500 000 scale.
- Detailed seismic zoning DSZ, was originally carried out for the most studied regions of perspective construction in 1:1 000 000, 1:500 000 scale or very rarely in 1: 200 000 scale.
- Seismic microzonation – SMZ, in 1:25 000 scale or greater, contained in engineer investigation system.

The results of seismic zoning have to be the appropriate map creation GSZ, DSZ and SMZ. DSZ differs from GSZ in investigation scale. At the same time, in DSZ process may and must be studied all potential sources of possible earthquakes, which may be not taken in account, e.g. they have relatively small seismic potential during GSZ analyzing. It has to be mentioned, that in the real conditions the consequences of seismic hazard generation with that types of sources may have, if not great, but noticeably negative effect. At the same time both types of zoning are very similar, nothing to say about minuteness.

The third stage or stage of seismic hazard assessment in SMZ type has absolutely other physical meaning, in spite of similar name with GSZ and DSZ. The SMZ using allows to take into account the seismic properties of site soils, including physicomechanical and dynamical properties of soil.

The SMZ map traditionally is a normative part of Building Codes, and regularly is revised. At the same time during the map design only huge geology-geophysical zones are taken into account, which the seismicity determined.

The assessment of seismic hazard of the site is carried out using necessity and probabilistic methods. The probabilistic analysis of seismic hazard assessment includes alternative models of seismic sources, the earthquake returne periods, the seismic signal attenuation and distance dependence, and much vagueness, caused by careless information of some parameters, and by random character of seismic events. In the necessity analysis of seismic hazard assessment the vagueness is not considered, only the extreme seismic effect is estimated on the real site, using near earthquake sources with fixed magnitudes. There are many domestic and foreign algorithms and programs for this purpose.

Practically all the previous maps of seismic zoning, from the first map in the former USSR till the last but one map were necessity. They not take into account the main characteristic of seismic regime of seism active territory, although in the middle of 40th S.V. Medvedev proposed to bring in seismic hazard zones internal differentiation including the strong earthquake return periods and assumed constructions durability. Then U.V. Riznichenko created algorithms and programs for seismic "shakeability" estimation. But all these progressive development of domestic seismologists, like their other ideas were not brought in use. then western countries begun to create seismic zoning map in exceeding (or nonexceeding) probability of seismic hazard in given times intervals.

The vagueness conditions, are always presented in nature, so the necessity method in the seismic zoning is incompetent. The seismic zoning process must use only probabilistic methods. The risk is always presented, but it must be estimated and reduced to minimum. These ideas are presented in the new more progressive maps of Russia general seismic zoning - GSZ -97. For the first time in Russia was proposed to use the probability map kit GSZ -97 for different constructions. General map GSZ -97 is presented on figure.

Wide spread usage of GSZ is caused by insufficient development of DSZ and distinct labor-intensiveness of its realization for researchers. Prof. Ulomov and his colleges use modern methods instead of ancient and out of date approach. In the same time the GSZ materials using sometime is impossible due impossibility to use more detailed information of regional and local materials including tectonical materials. The map generalization is enough for state overall planning, but is not enough for reliable estimation of real objects seismic conditions.

The process of Detailed seismic zonation is very complicated and expensive complex of geology tectonical, geophysical and seismical investigation for quantitative estimation of seismic effect in any site of perspective region.

That type of investigation consists of all methods used in DSZ, but estimated quantitatively the source (background) seismic effects only on concerned site GSZ (more precisely for mean soil conditions or 2nd seismic category soils on site).

So, it is necessary to develop DSZ approach. The modern DSZ has clear and argumented content. There is huge Strong Motion Data Base with many records of soil velocity and acceleration, including South Caucasus Countries. Now, there are many modern computer programs, reliable digital velocity and acceleration registrators, now we may obtain many records of earthquakes. So, it is possible to realize DSZ purpose using reliable data. And, in spite of updating initial seismicity (UIS) for DSZ we have tye possibility to estimate site seismic hazard.

It must be told, that UIS-DSZ methodic always formed parallel with GSZ methodic, but the scale differs, and some additional methods. There are some methods that may be used in GSZ and DSZ for seismic generic structures (SGS) identifications, it is identification of zones of danger earthquake appearance.

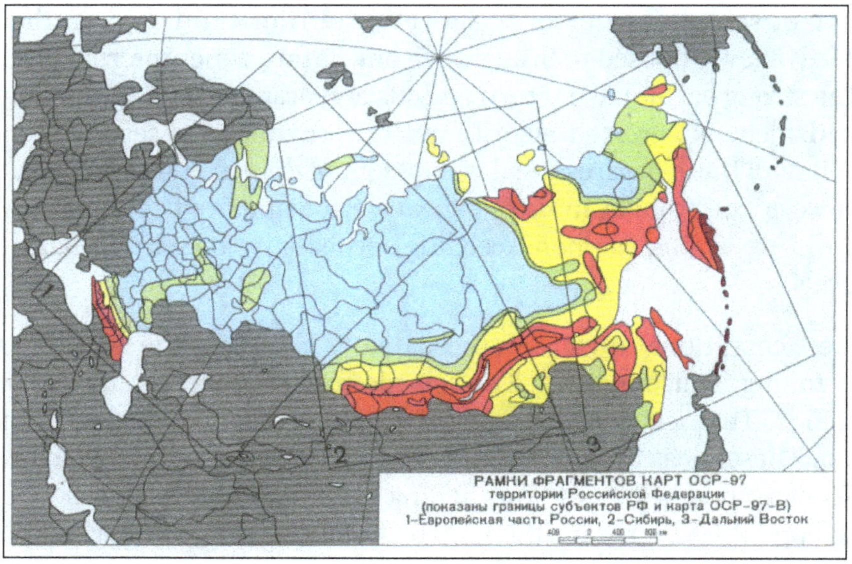

Map of General seismic zonation GSZ-97 of Russia.

Seismogeological Method

Using the first epicentral zone investigation in the end of XIX and beginning of XX centuries Abich G. and Lagorio A.E. find out the dependence between earthquake and tectonic structures. Mushketov I.V. writing about Vern earthquake 1887 year, told that Turkestan earthquake is connected with discontinuous disturbance. He wrote earthquakes "culminate on the boundary of the most huge and new disturbance". Besides, he wrote that some groups of earthquakes are connected with lines, transversal to common stretch of rugosity, e.g. connected them with transversal structures in modern terminology. K.I. Bogdanovich analyzing Kebi earthquake consequences in the North Tien Shan, introduced new term – seismotectonic element, and for the first time proved the seismic shock migration inside seism active zone.

So, seismogeological method was able to connect strong earthquakes with tectonic structures. Those bonds later were named as geological seismicity criterion and were used in other methods.

Seismotectonical Methods

Seismotectonical method was introduced in the end of 40th years of XX century by Gubin I.E. when investigate Garm region on the Pamirs - Tien Shan border. He connected earthquakes with discontinuous zones some tenth km wide. He wrote, that "seismogenity degree" is stable all over the zone, "seismicity degree" may be ascribed to other similar zones, "if this structures, using geological data, are connected by mutual evolution process with equal intensity". This method (Gubin seismotectonics law) says, that in a given geological medium in the active structures of the same type and size, maximum earthquakes, originate from the rock displacement along the active rupture, have equal magnitudes and sources. Seismotectonical method accents on geological seismicity criterion – the velocity of young rupture displacement.

Seismostructural Method

Seismostructural method developed in the mid-50's, by V. Belousov, A. Goriachev, I. Kirillova, B. Petrushevsky, I.A. Rezanov, A.A. Sorsky, but most fully reflected in works of B. Petrushevsky. Earthquakes associated with large structural complexes-blocks allocated by using the historical-structural analysis and discontinuous joints.

Large-scale analysis blocks allowed to associate with them (and the underlying faults) varied range of depths of earthquakes (most profound on the articulation of the Pacific with Eurasian and the American continents). Picture of the strong earthquakes focuses with different three-dimensional structures of the Earth's crust was further developed in the works of G.P. gorshov. However, this promising direction needs to be fleshed out.

Tektonophysical Method

Tektonophysical method developed in the second half of the 50-ies by M.V. Gzovsky. The method connects the earthquake with the maximum tangential stresses area, which is in conjunction with the maximum gradients of average speeds of tectonic movements and breaks. The energy of the earthquake was put by M.V. Gzovsky in dependence on a number of factors. But a precise calculation is impossible because the mechanical properties of the Earth's crust and its viscosity in Maxima tangential stresses can be evaluated only in qualitative terms.

Method of Allocating Quasihomogeneous Zones

In the late 50-ies started to be developed method of allocating quasihomogeneous zones of earthquakes for one or all geological and geophysical criteria, some of which have tectonical nature. However, these criteria have not been effective in a number of regions.

Since the choice of number and encoding parameters and their combinations are endless, equally infinite may be variants of map M_{max}. In connection with this were analysed practically all existing geologic-structural, seismic and geophysical maps for the territorial zoning using seismotectonic capacity (combined geological criteria reflecting the characteristics of the medium properties and the intensity of tectonic process). Based on mathematical patterns is forecasting of magnitude M_{max} with reference site to the rest territory. The approach developed by Reisner and Ioganson, where reference sites were used in all of the zones of the planet. The analysis involved areas with variety of tectonic properties, where the seismicity criteria are mixed. Naturally, the common criteria were often not the fault criteria (thickness of the Earth crust, the heat flux density, height, isostatic gravity anomalies, the depth of the consolidated Foundation, etc.) The method later became known as the "extraregional method".

Method of Seismoactive Nodes

Structural refinement of earthquakes has allowed to V.M.Reiman at the turn of the 50 's and 60's to make an idea using the Central Asian material of disjunctive nodes in which the strong earthquakes concentrate or sejsmogenetic nodes. Later became actively used the term the seismically active sites. The best method was developed by E.Y. rantsman, which extended the scheme to many orogenetic regions of the world. E.Y. Rantsman links with sites the earthquakes epicenters, stressing that "the earthquakes focuses can reach hundreds of miles away and go far beyond the morfostructural nodes". To classify the structures of seismicity was proposed the complex system of formalized criteria (distance from the edges of the site, type of terrain, maximum height, and area of friable deposits) and mathematical apparatus. Study of seismogenerating structures made it possible to include cross rises in the number of structures that make up the nodes.

Paleoseismological Method

Paleoseismological method allows using paleoseismodislocations to trace possible seismic sources zones (PSS zones) and estimate their magnitude and seismic intensity. To evaluate these parameters the seismotectonic dislocation is used. Currently, there are many formulas (public and regional), describing the statistical associations between seismodislocations (length, amplitude displacement) and seismological parameters (magnitude, the depth of the epicenter, intensity of seismic vibrations) of earthquake. To determine the occurrence frequency of earthquakes it is necessary to have a reliable assessment of the age of paleo-seismic dislocations. All possible approaches are used: geological-geomorphological, archaeological, historical data and radiocarbon dating of sediments, broken by seismodislocation and later. Dendrohronological method is used, which takes into account the changes in the growth of trees associated with earthquakes, as well as lihenometrical method of dating seismogenical samples and dedicated to its some species of lichens.

Detailed Seismic Zoning

As an example, the assessment of seismic hazard, let's consider some estimations on the DSZ level in the territory of North Ossetia. On the basis of an analysis of methods for identification of PSS zones was elected out of regional seismotektonic method to objectively identify the sejsmogenic sources. Despite some shortcomings, the entire method is characterized by the quantitative indicators and has strong decision-making apparatus. The method has been used by prof. E.A. Rogozhin in North Ossetia when solving various scientific tasks. In addition, using this method some similar tasks were solved not only for Russian but also overseas territories (Israel, Italy etc.). At the same time, this does not preclude obtaining reliable results and other known methods.

The methodology used in most probabilistic seismic hazard analysis was first defined by Cornell and as usually accepted it consists of four steps: 1. Earthquake source zones (SSZ), 2. Recurrence characteristics for each source, 3. Estimation of earthquake effect and 4. Determination of hazard at the site.

Earthquake Sources

As a rule, today probabilistic assessment of seismic hazard is used all over the world for the identification of seismic loads for the engineering projects. The probabilistic approach is a more systematized method for the assessment of quantity, sizes and location of future earthquakes than any other methods. Formal procedures for the probabilistic assessment include the determinations of spatio-temporal ambiguities for the expected (future) earthquakes. The computer program EQRISK of McGuire became the main stage in the method development. The program became widespread and is very popular up to present day. In this connection the probabilistic assessment of seismic hazard is often called Cornell McGuire's method. The program includes integration on ambiguities distribution.

The Caucasian region is characterized by high intensity of dynamic geological processes and hazards, connected with them, of both natural and technogenic character. The most clearly expressed among these hazards is seismicity, which is accompanied with wide range of secondary processes. Earth surface ruptures, activation of known earlier inactive faults, landslip phenomenon, collapses, avalanches, creep and subsidence of the earth surface, activation of surface structures, soil liquefaction and other hazardous phenomena can be noted among them.

The investigations on determination and parameterization of the seismic source zones in recent decades has been realized by V.P.Solonenko, V.S.Khromovskikh, E.A.Rogozhin, V.I.Ulomov, V.G.Trifonov, I.P.Gamkrelidze and others.

On basis of the results of the active faults study located southward of the Great Caucasian ridge, parameters of seismic source zones were chosen according to data of I.P. and to the north of the ridge they were chosen on data. According to the results of the executed expert evaluation of seismic potential (M_{max}) the maps of seismic sources zoning of the territory of North Ossetia (zones of possible seismic sources - PSS zones) were made up.

A new original method of more accurate ascertainment of the boundaries of seismogenic source (fault) active part and assessment of the potential of seismic source hazard (at works of detailed seismic zoning – DSZ) has been worked out in recent years.

Let's consider the process of territory seismic hazard assessment for explanation of procedure usage by the example of the Central Caucasus (the territory of the Republic of North Ossetia-Alania).

PSS zones are referred to the active fault systems, singled out on a basis of interpretation of the materials of remote sensing and geological data. Decoding of multispectral three-channel space images of Landsat–4/5 (resolution 30 m) and Landsat–7 (resolution 15 m) was realized. Decoding of space satellite photos was executed in colored multispectral variant as well as in black-and-white variant. Different variants of the image synthesis were used for the analysis of polyzonal scanner pictures. Besides, identification of the lineaments was also executed separately on channels. Combined deductive – inductive approach was used for lineaments identification: integrated structures were decoded on the base of strongly generalized images with the following zooming in for detailing and vice versa local peculiarities of tectonic and exogenous structures with the following zooming out and generalization. The method of stepwise generalization was used with quantization on the scale levels 1:25000; 1:50000; 1:100000; 1:200000; 1:300000; 1:400000; 1:500000. In the scale range 1:25000 - 1:1 500000 space photomap on basis of snapshots Landsat-7 is used and in the range 1:500000-1:2 millions – space photomap, created on basis of Landsat–4/5 snapshots.

Extensive lineaments systems were identified with known faults, which were qualified on modern stage as active. The name of PSS zones was formulated on basis of faults and large settlements names. Morpho-kinematics of active faults is the base for qualification of seismic displacements kinematics in PSS zones. Hypocenters depth of expected earthquakes was calculated from the depth of fault plans, the depth on geophysical anomalies data and from the magnitude of expected events.

Maximum magnitude of expected earthquakes (seismic potential, M_{max}) was assessed on the results of usage of the over-regional seismotectonic method of seismic hazard assessment.

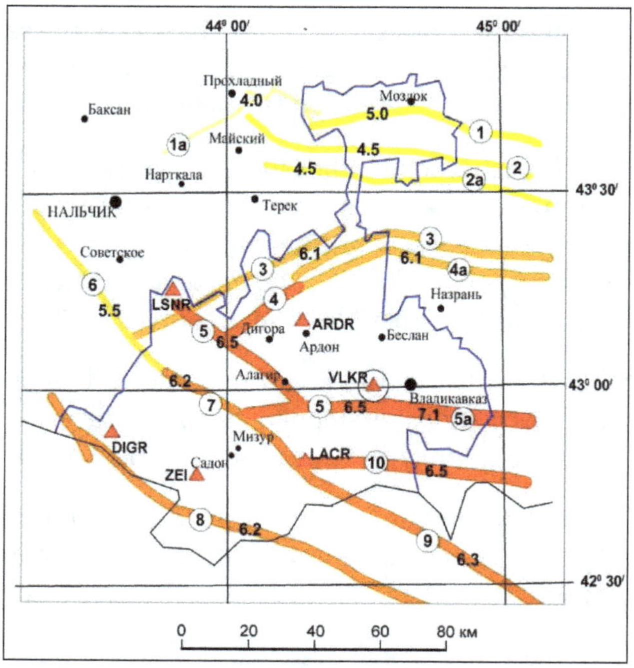

Map of PSS zones of the territory of the Republic North Ossetia-Alania. Red triangles – basic seismic stations in the region. Blue and black lines are the state borders of North Ossetia.

In 2007 it was determined on data of field investigations that for the urbanized territories of North Ossetia the most hazardous are Vladikavkaz, Mozdok, Sunzha and Tersk PSS zones.Parameterization of seismic sources was made after creation of these maps, i.e. maximum possible magnitude M_{max} for each seismic source was assessed. This is the most difficult problem in the process of parameterization of PSS zones. M_{max} was determined on the data of a number of authors.

The second essential parameter, which characterizes expected earthquakes, is sources depth range, where the majority of seismic events with corresponding magnitude generate. According to the numerous investigations, Caucasus is the region with upper crust part location of seismic sources – their depth doesn't exceed 20–25 km (deeper seismicity is observed in Tersk-Sunzha zone in the area of Grozniy city and in Caspian Sea). As sources distribution on depth for this region wasn't executed, average value of depth (equal to 10 km) was taken for calculations.

Table: PSS zones for North Ossetia characteristics.

No.	PSS zone	Magnitude	H, km	Kinematics.
1	Mozdok eastern	5.0	10	Reverse faulting
1a	Mozdok western	4.0	5	Strike-slip
2	Tersk	4.5	5	Reverse faulting
3	Sunzha northern	6.1	15	Reverse faulting
4	Sunzha southern (western branch)	6.5	15	Strike-slip
4a	Sunzha southern (eastern branch)	6.1	15	Reverse faulting
5	Vladikavkaz (western branch)	6.5	15	Reverse faulting
5a	Vladikavkaz (eastern branch)	7.1	20	Reverse faulting
6	Nalchik	5.5	10	Strike-slip
7	Mizur	6.2	15	Strike-slip
8	Main ridge	6.2	15	Reverse faulting
9	Side ridge	6.3	15	Reverse faulting
10	Karmadon	6.5	15	Reverse faulting

Reoccurrence Characteristics

For the assessment of ratio parameters between reiterations during the process of execution of a number of investigations on the international projects the earthquake catalogue was checked and specified.

Corrected Catalogue of Caucasus contains data for more than 61000 of earthquakes, including 300 historical events, which happened during 2000 years. This catalogue was checked and corrected. Some hypocentral parameters of earthquakes were recalculated.

Threshold of magnitude for the whole catalogue and a and b values of the frequency-magnitude law were determined for large tectonic zones, as their calculation for certain PSS zones was impossible because of data absence. Value of b of the frequency-magnitude law is determined by formula of Gutenberg-Richter:

$$\lg(N/T) = a - bM,$$

where a and b are parameters, the inclination and level of recurrence graph at M=0.

For each PSS zone (both linear and square) frequency of earthquake origination was studied on basis of observed seismicity. For study of Gutenberg-Richter ratio earthquakes were referred to the separate faults or PSS zones taking into account accuracy in epicenter determination. Because of the shortage of data about accuracy of location determination average model was accepted. This model supposes that mistakes have normal distribution with standard deviation equal to 3–4 km. Distances from each event to the all PSS zones were measured and only zones, which were on closer distances from the event than three standard deviations, were taken into account. Based on distances value, weighting coefficient was assigned to each zone, from the curve of density distribution of the standard deviation possibility.

Estimation of Earthquake Effect

Earthquake effect was estimated using two different parameters: macroseismic intensity and peak ground acceleration (PGA). Macroseismic intensity (MSK scale) was traditionally used for seismic zonation in former USSR. Macroseismic and instrumental data on 43 significant earthquakes occurred in Caucasus were revised to obtain the necessary information. Data on 37 earthquakes was selected and in some cases were compiled new isoseismal maps in the 1:500 000 scale. In a process of computations was observed a fact that the value of the attenuation coefficient in vicinity (within the limits of the first three isoseismals) of the source of the M_s>6 earthquake is very high (˜4.5-5.0), in comparison with small and moderate events (~3.4). This fact has been tested on the other Caucasian strong earthquakes (M_s>6) and in general has been confirmed. In spite of the lack of data in the first approximation the equation of correlation in this case obtains the following form for small earthquakes:

$$I = 1.5M_s - 3.4\lg(\Delta^2 + h^2)^{1/2} + 3.0,$$

and

$$I = 1.5M_s - 4.7\lg(\Delta^2 + h^2)^{1/2} + 4.0,$$

for large events.

It should be noted, that for hazard estimation we have used the second relationship. Besides that we have restricted maximal value in epicentral area for M=7, (6.5) earthquakes with intensity 9, M=6 (5.5) earthquakes with intensity 8, etc. this was done to avoid very high intensities in epicentral area. The epicentral areas were estimated using relationships for earthquake source sizes given.

On the other hand strong motion instrumental data in Caucasus and adjacent regions allows us to use PGA and spectral acceleration attenuation law for seismic hazard analysis. Since the installation of the first digital strong-motion station in the Caucasus area 451 acceleration time histories from 269 earthquakes were recorded. Based on the acceleration time histories recorded between June 1990 and September 1998 with the permanent and temporary digital strong-motion network in the Caucasus and adjacent area, 84 corrected horizontal acceleration time histories and response spectra from 26 earthquakes with magnitudes between 4.0 and 7.1 were selected and compiled into a new dataset. All time histories were recorded at sites where the local geology is classified as "alluvium". Therefore the attenuation relations derived in this study are only valid for the prediction of the ground motion at "alluvium" sites.

The calculation of the correlation coefficients and the residual root mean square was performed with the well known Joyner and Boore two step regression model. This method allows a de-coupling of the determination of the magnitude dependence from the determination of the distance dependence of the attenuation of ground motion. Using the larger horizontal component for spectra of the selected acceleration time histories, the values of coefficients were obtained for the coefficients at different frequencies. Because it is easy to obtain peak acceleration from corrected acceleration time histories, empirical attenuation models with peak ground acceleration as dependent parameter have always played an important role in different seismic hazard and earthquake engineering studies. The resulting equation for larger horizontal values of peak horizontal acceleration is:

$$\text{Log}\,PHA = 0.72 + 0.44M - \log R - 0.00231 + 0.28p,$$

$R = (D2 + 4.52)^{1/2}$, where PHA is the peak horizontal acceleration in [cm/sec^2], M is the surface-wave magnitude and D is the hypocentral-distance in km. p is 0 for 50-percentile values and 1 for 84-percentile.

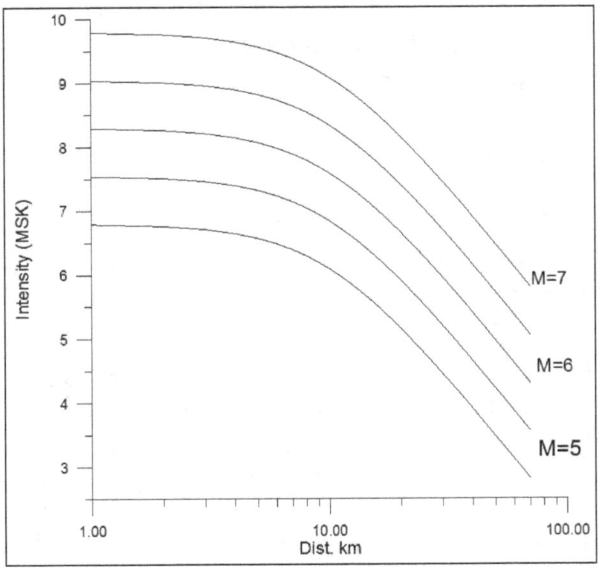

Attenuation model for intensity (MSK).

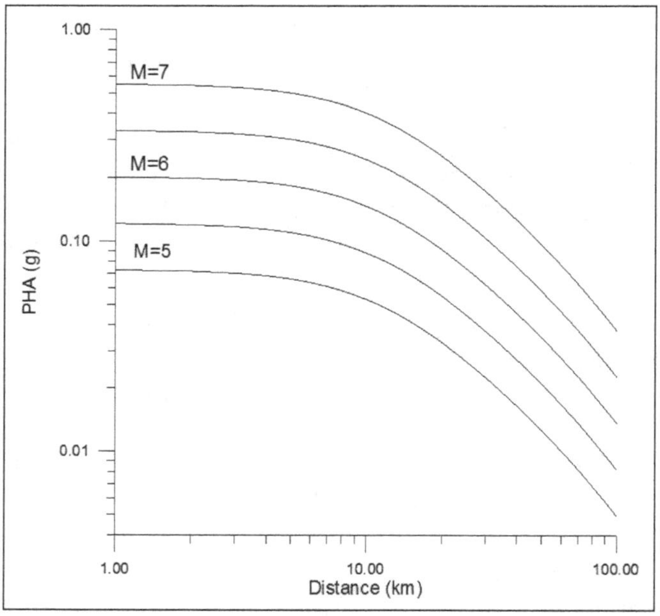

Attenuation model for acceleration.

It is important to bear in mind that all equations given above represent a best fit of the selected dataset, and therefore represent mean values about which there is a considerable scatter. In the case of the attenuation model for the larger horizontal value of the peak horizontal acceleration the predicted mean plus one standard deviation is equal to 1.91 times the mean value. The scatter of the pha-models is the same as similar models for Europe and Western North-America.

The comparison of the attenuation relationships for peak horizontal acceleration with similar relations for other areas shows a good agreement with the models from Western North-America. It is obvious, that the attenuation in Europe is lower compared to the Caucasus and adjacent area. The predicted peak values in the near-field are higher than the corresponding values obtained with other European models.

Determination of Hazard

The probabilistic seismic hazard maps (the maps of detailed seismic zoning) have been constructed for the total area of North Ossetia in scale 1:200000 with exceedance probability for a period of 50 years (standard time of building or construction durability!) with 1%, 2%, 5%, 10% in GIS technologies, which corresponds to reoccurance of maximum probable earthquake for a period of 5000, 2500, 1000 and 500 years. The longer the period of time the higher the level of possible intensity. For a period of 500 years only a small part will be occupied by the zone of 7 intensity earthquake, for a period of 1000 years – 8 intensity and at 2500 years 9 intensity earthquake appearance, correspondingly.

Cornell approach, namely computer program SEISRisk- 3, developed in 1987 by Bender and Perkins was used for the calculations. The map of observed maximum intensity was compared with the maps of different periods of exposition and the most real map was chosen on a basis of the analysis of differences between the observed and calculated maps. According to these criteria the map of 5% probability with exceedance probability of 50 years can be recommended for seismic zoning of the territory of North Ossetia. Besides, for the first time probability maps of seismic hazard for Russian territory were made in acceleration units in scale 1:200 000 with exceedance probability for a period of 50 years - 1%, 2%, 5%, 10%.

According to the conception, it is necessary to use the data, which is maximum approximate to the real engineering-geological conditions, at assessments of territory seismic hazard. For the territory of North Ossetia the exposition equal to 1000 years is the most approximate to real conditions for mass building. It is necessary to consider greater exposition, for example, 2500 years etc. for unique buildings and constructions.

The maps of 5% probability are likely to be used for the large scale building, i.e. the major type of constructions, whereas the maps of 2% probability should be used for high responsibility construction only. One can see great hazard in the south of North Ossetia on the map, where exists the increased level of seismic hazard (due to powerful Vladikavkaz fault, lying nearby).

As a matter of principle it is possible to make maps in scale 1:100 000 etc., but it actually makes no practical sense. Although accuracy of such maps must be higher, adequacy of the results can be considered as doubtful due to absence of reliable data on local peculiarities of past, i.e. historical earthquakes display. Laboriousness (irretrievable) at that increases multiply.

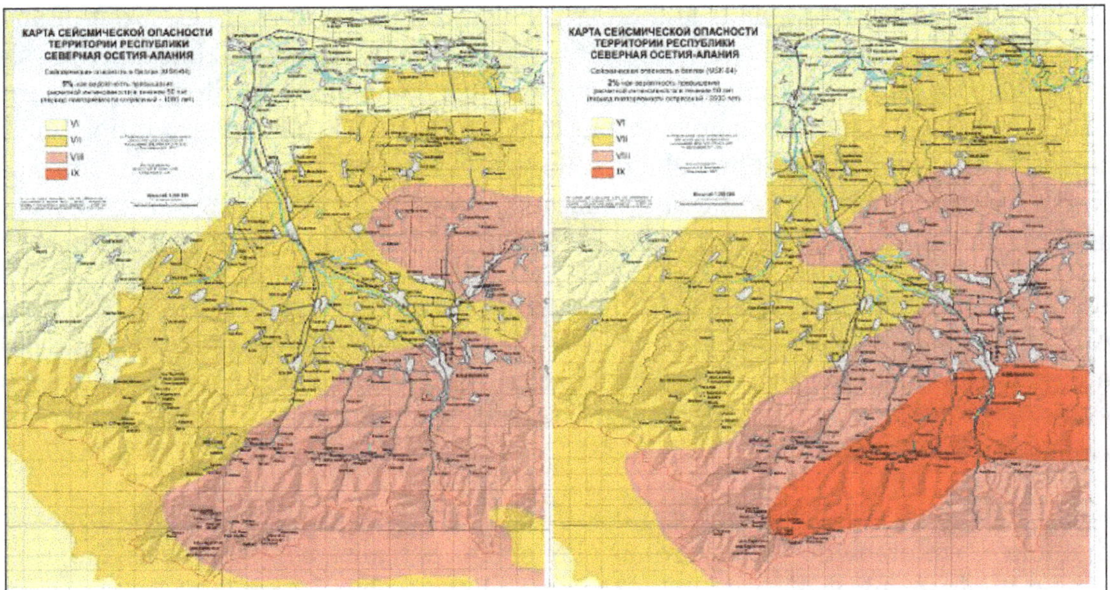

a) b)

Probabilistic maps of seismic hazard (DSZ) in the intensities (MSK-64) with the exceedance probability 5% (a) и 2% (b) for North Ossetia territory and adjacent areas.

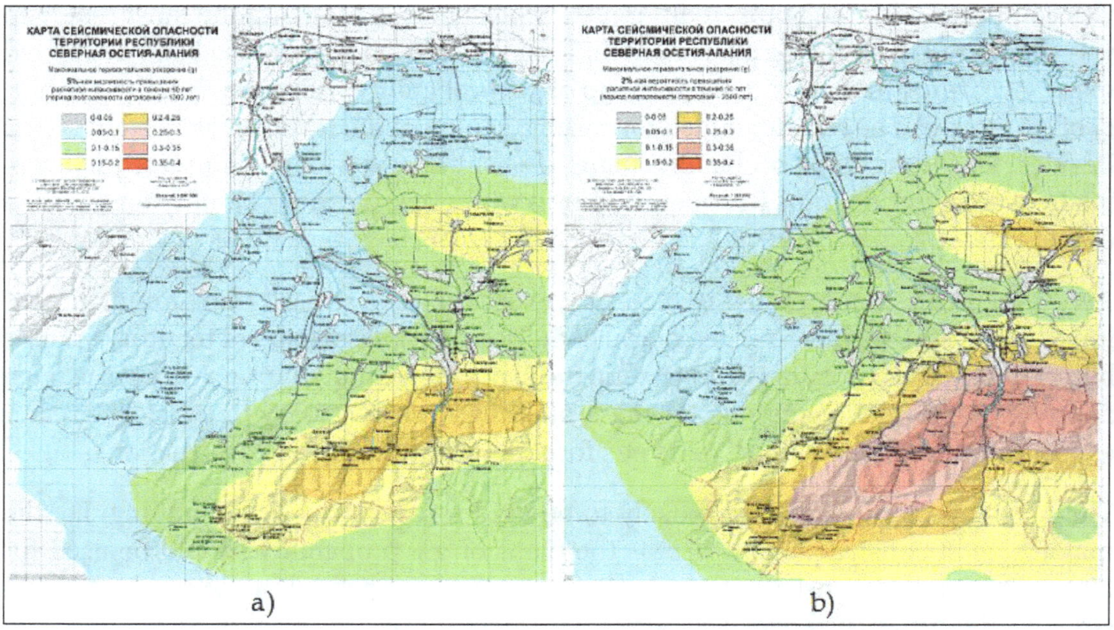

Probabilistic map of seismic hazard (DSZ) in accelerations (PGA) with exceedance probability 5% (a) and 2% (b) for North Ossetia territory.

The scientists from Vladikavkaz in collaboration with the colleagues from the Institute of Physics of the Earth of RAS not only offer to use large-scale maps but also decided to continue investigations and cover the whole Northern Caucasus in scale 1:200 000. So, maps of seismic hazard can be made up in scale 1:200 000 for the Republics of Chechnya, Ingushetia, Kabardino-Balkaria, Stavropol and Krasnodar areas and the other territories. Taking into account, that faults and other peculiarities of the territory exist out of any boundaries, including state boundaries, it is possible to make unusual but quite physically proved single general map of detailed seismic zoning of the territory of Northern Caucasus in scale 1:200000, moreover, one can make them for different exposition times and accordingly for different probabilities. So, created maps of detailed seismic zoning of North Ossetia conform to earthquake realization once in 500 years, 5% - in 1000 years and 2% - in 2500 years. The level of seismic hazard grows with the time increase etc.

It is possible to make detailed maps of seismic hazard for the whole Caucasus, including Azerbaijan, Armenia and Georgia, due to the features of spreading of hazardous seismic sources, which "neglect" states' boundaries. It is also possible to develop the maps jointly with Turkey and Iran and it's real to include such countries as Israel, Egypt, and Lebanon etc.

Maps of detailed seismic zoning can be called "long-term" prediction maps. It means that long-term prediction of hazardous phenomena is realized on their basis and, correspondingly the place of earthquake-proof building-stock is determined.

Essentially, the long-term maps of expected intensities locations are that of described maps of detailed seismic zoning. Indeed, that evacuate people from the hazardous territory before expected earthquake is impossible, but it is real to prevent population burring under destroyed or, to be more precise, differently damaged buildings, which is formed on basis of such maps. The more educated society is the less seismic risk, i.e. economic and social losses. So, the priorities are clear.

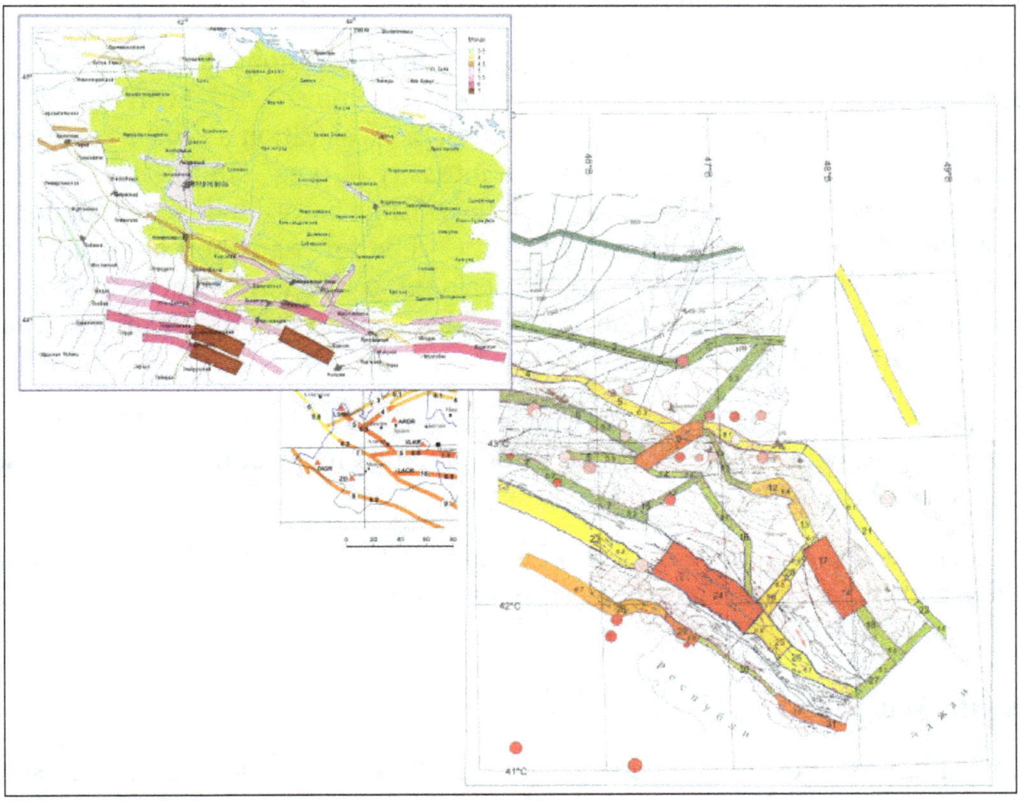

The mosaic of maps of hazardous potential seismic sources on the territories of the Northern Caucasus.

On basis of the given maps it is necessary to make up the maps of seismic microzonation (SMZ) of cities and large settlements of each certain subject of the Russian Federation with the usage of the most modern standard methods and tools, but in scale 1:10 000. The probabilistic maps of SMZ were first developed in the Center of Geophysical Investigations of Vladikavkaz Scientific Center RAS and RNO-A. Such maps of SMZ are direct and reliable base of earthquake-proof design and object construction.

Besides, it is necessary to note that at usage of the traditional units of macroseismic intensity the boundaries between different zones are characterized by sharp changes, which obviously do not correspond to the real situation of monotonous change of intensity for homogenous soil conditions of the investigated territory. No doubt, it will form evident inaccuracies at the assessment of the level of seismic hazard of this or that territory. The practical usage of artificial intensity subdivision, for example, in the form of 7.2 or 8.3 points is not validated enough from the theoretical point of view.

So, firstly, it is not usually explained how these fractional assessments are obtained and, secondly, the following transition to the acceleration units (obviously, according to foreign data, as there are no acceleration records for forming reliable correlation in Russia), undoubtedly, forms considerable inaccuracy and it is hardly ever physically proved because of the formality of the parameter of "intensity" itself.

On the other hand, at seismic influence assessment at earthquake - proof design engineers use the acceleration values, (strictly speaking, conveniently) corresponding to specified intensities. Thus, it's assumed that design acceleration a = 0.1 g corresponds to the intensity 7 earthquake, 0.2g – to the intensity 8, 0.4g – to the intensity 9 etc. Namely it makes possible to design maps of the seismic hazard independently in units of PGA. Such maps for the territory of North Ossetia for exposition of 50 years with exceedance probability 1%, 2%, 5%, 10% in scale 1:200 000 were created. It is obvious that at changing of smoothering step it is possible to obtain smooth variations of accelerations directly used as design impacts.

In contrast to the maps of general seismic zoning (GSZ) with a scale of M 1: 8000000 and, at the best, with the scale M 1:2500000 obtained maps of both types on a scale 1:200000 can be referred to the DSZ type maps.

Thus, these materials allow assessing seismic hazard on a detailed level, according to the known formulas to calculate the macroseismic field of seismic effects on a scale that may provide a reliable basis for SMZ.

Seismic Microzonation of Territory

Seismic microzonation (SMZ) actually is final stage of seismic hazard assessment. SMZ results are direct foundation for earthquake-proof construction. In the process of seismic microzonation sites with etalon ground conditions corresponding to specified seismic hazard level are specified. In Russia grounds with mean seismic properties for given territory are traditionally referred as etalon ground conditions. Usually these are soils with shear wave velocity of 250–700 m/s. In Georgia, for example, in dependence of specific engineering-geological situation etalon grounds in their seismic properties can be worst or mean for given territory. In USA firm rock grounds are referred as etalon. Seismic microzonation consists in intensity increments calculation caused by differences in ground conditions. Works on seismic microzonation are realized by instrumental and calculational methods.

Instrumental Method of Seismic Microzonation

Instrumental method is the main SMZ method. Exactly it urges to solve a problem of forming earthquake intensity forecast. At the same time the calculation method, which allows to model any definite conditions of area and influence features, is often characterized by more reliability. It has great importance to soil thickness with high power. Combined usage of both methods significantly increases results validity.

Seismic Microzonation on the basis of Strong Earthquakes Instrumental Records

It is supposed at usage of strong earthquakes records for SMZ purposes, that at some strong seismic influence the observing soil behavior is adequate to the display of their potential seismic hazard at future strong earthquakes. This fact was the reason of stimulation of a number of large international scientific-research projects on organization of long-term instrumental observations with the help of powerful measurement systems in the Earth's different regions with high seismic activity for the purpose of obtaining the strong movements of soils, which are the base of buildings and constructions (the groups SMART-1 and SMART-2 on the Taiwan island etc.)

At the same time, presence of unit record of a real strong seismic influence at its inestimable value for SMZ often can't give the adequate forecast of soil behavior at a next following strong earthquake. This problem can be solved by creation of a number of records of seismic influences, generated by hazardous for the zoned territory active fractures, i.e. by zones of possible earthquake source (PES).

Seismic Microzanation with the help of Weak Earthquakes Records

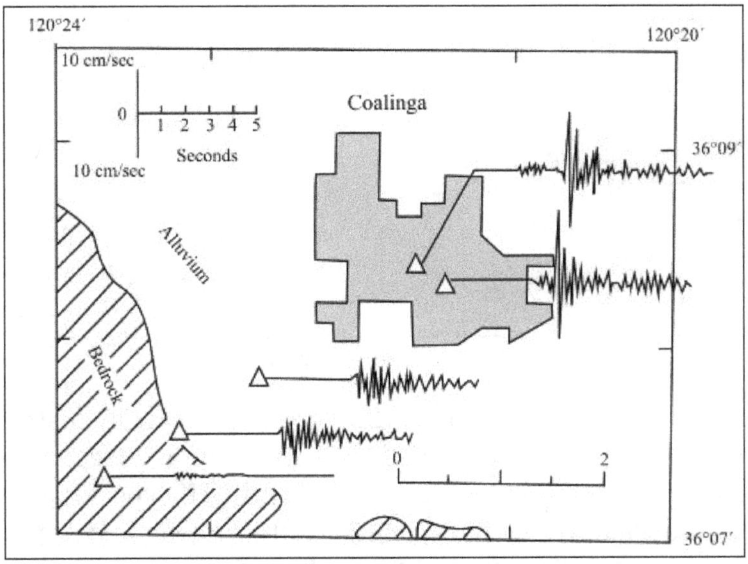

Scheme of California earthquake in Koaling sity.

In the connection of the fact that strong earthquakes occur seldom, the intensity increments, as a rule, are assessed by records of weak earthquakes, when a linear dependence between the dynamic stress and the deformation takes place.

Soil conditions considerably change the right shape of the original undistorted signal, incident from the crystal foundation. Complex shapes of isoseisms pointed out to the undoubtful link between the earthquake display intensity and soil conditions.

Increase of the soil thickness depth (alluvium) considerably changes the character of earthquake records in the process of approaching the city.

Calculation of intensity increment with the help of weak earthquakes is realized by the formula:

$$\Delta I = 3{,}3 \lg A_i / A_0,$$

where,

A_i,A_o are the amplitudes of investigated and etalon soils vibrations. The usage of tool in the form of registration of strong and weak earthquakes needs the organization of instrumental observations in a waiting mode.

Seismic Microzonation using Microseisms

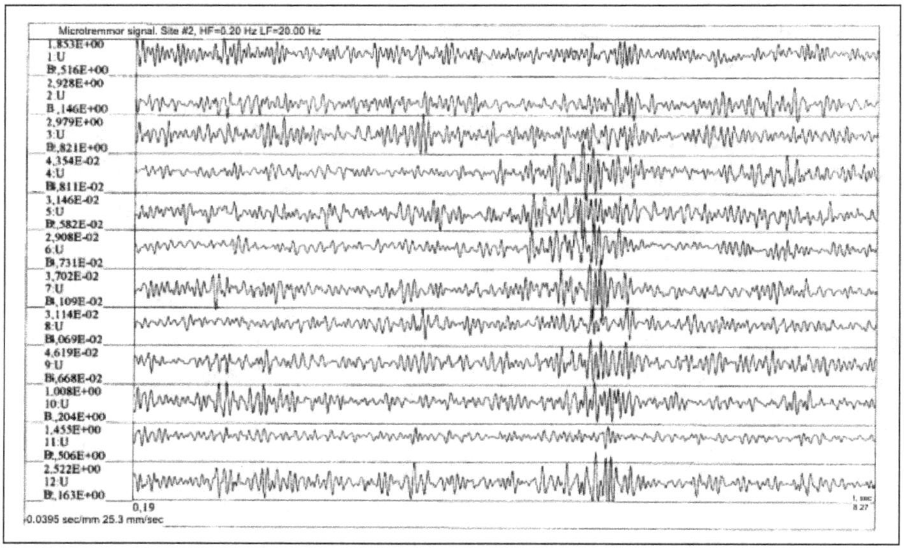

Microseisms records.

The results of microseisms observations are used as subsidiary instrumental tool of SMZ. Predominant periods are determined at that in order to assess resonance properties of soils and amplitude level of microvibrations. Strictly speaking, the reference of microseism on their origin to the purely natural phenomena is not quite correct. Numerous artificial sources, influence degree of which can't be controlled, undoubtedly, take part in their forming along with the natural sources.

Intensity increment for strong earthquakes on microseism is calculated by the formula:

$$\Delta I = 2 \lg A_i / A_0,$$

where

A_i,A_o are the maximum amplitudes of microvibrations for investigated and etalon soils.

Impossibility of the compliance of necessary standard conditions of microseism registration and large spread in values of maximum amplitudes limit the usage of microseism for calculation of soil intensity increment. The above mentioned causes the application of microseism tool only in complex with other instrumental tools. Spectral features for different sites are estimated by means of H/V-rations.

Seismic Microzonation using Explosive Impact

The intensity increment ΔI of the soils of the zoned territory is calculated by the formula at usage of weaker explosions:

$$\Delta I = 3{,}3 \lg A_i / A_0,$$

where A_i, A_0 are vibrational amplitudes of the investigated and etalon soils.

Execution of powerful explosions on the territory of cities, settlements or near the responsible buildings is connected with large and often insurmountable obstacles (technical and ecological problems, safety problems, labouriousness and economical expediency) and practically isn't used nowadays. This leads to the wide spreading of nonexplosive vibration sources.

Seismic Microzonation using Nonexplosive Impulse Impact

The features of SMZ methods development led to the situation when the tool of elastic wave excitation with the help of low-powered sources (for example, hammer impact with m = 8–10 kilograms) has become the most wide spread in the CIS countries, in order to determine S- and P-wave propagation velocities in soils of the typical areas of territory. Velocity values are used in order to calculate the intensity increment using the tool of seismic rigidities.

where $\rho_0 V_0$ and $\rho_i V_i$ is the product of the soil consistency and P-wave (S-wave) velocity – seismic rigidities of the etalon and the investigated soil accordingly.

The intensity increment, caused by soil watering, is calculated by the formula:

$$\Delta I = K e^{-0{,}04 h_{GL}^2}$$

where K = 1 for clay and sandy soils; K = 0,5 for large-fragmental soils (with sandy-argillaceous filler not less than 30%) and strongly weathered rocks; K = 0 for large-fragmental firm soils consisting of magmatic rocks (with sandy-argillaceous filler up to 30%) and weakly weathered rocks; h_{GL} is the groundwater level.

The simplicity and immediacy of practical application of S.V.Medvedevs' tool, which is called the tool of the "intensities", led to its widespread in CIS countries and countries of Eastern Europe, Italy, USA, India, and Chile in 1970-es. The tool of the "intensities" was advantageously different from other tools by the immediacy, simplicity in initial data obtaining and its processing and independence from seismic regime of the territory. It to a certain extent hampered the development and making up of new tools. Unfortunately, the calculation results of predicted values of intensity increment are often quite incorrect as data of macroseismic observations of destructive earthquake consequences shows.

By means of the special investigations it was determined that the reliability of calculated intensity increments considerably increases at usage of modern powerful impulsive energy sources.

The lowering of final results quality is to a certain extent caused by the fact that in the tool of "intensities" the seismic effect dependence in soils on frequency or "frequency discrimination" of soils and also the origin of typical "nonlinear effects" at strong movements isn't taken into account. A.B.Maksimov tried to remedy this deficiency by developing the tool, where frequency peculiarities of soils were taken into account:

$$\Delta I = 0.8 \lg \rho_0 V_0 f_0^2 / \rho_i V_i f_i^2$$

where f_o, f_i are predominant frequencies of etalon and investigated soils.

A.B.Maksimovs' tool didn't find wide distribution, as frequency differences of soil vibrations with sharply different strength properties (at usage of traditional for the seismic exploration of small depths low-powered sources) were insignificant and the calculation results on the formulas $\Delta I = Ke^{-0{,}04h_{GL}^2}$ and $\ddot{A}I = 0.8\lg \tilde{n}_o V_o f_{wao}^2 / \tilde{n}_i V_i f_{wai}^2$ were practically similar.

Intensity increment was determined by the following formula:

$$\ddot{A}I = 0.8\lg \tilde{n}_o V_o f_{wao}^2 / \tilde{n}_i V_i f_{wai}^2$$

where f_{wao}, f_{wai} are weight-average vibration frequencies of etalon and investigated soils.

Weight-average vibration frequency of soils was calculated at that on the formula:

$$f_{\tilde{n}\hat{a}} = \Sigma A_i f_i / \Sigma A_i$$

where A_i and f_i are the amplitude and the corresponding frequency of vibration spectrum.

Surficial gasodinamical pulse source (SI-32).

Seismic Microzonation using Vibration Impact

At usage of a vibration source the calculation of intensity increment is realized with the help of the formula:

$$\Delta I = 2\lg S_i / S_0,$$

where S_i and S_o are the squares of vibration spectra of investigated and etalon soils.

The developed tool was used at SMZ of the territories of cities Tbilisi, Kutaisi, Tkibuli, single areas of the Bolshoy Sochi city. The tools' feature consists in the fact that it allows to assess soil seismic hazard without any preliminary investigations: at realization of direct measurements of soil thickness response on standard (vibration or impulse) influence. Later the formula was successfully used at SMZ of the sites of Novovoronezh Nuclear power-plant (NPP) with the help of an impulsive source.

Vibration source (SV-10/100).

Seismic Microzonation on basis of taking into Account Soil Nonlinear Properties

The comparison of the absorption and nonlinearity indices with the corresponding spectra of soil vibrations shows that at higher absorption the spectrum square prevails in LF field and at high nonlinearity it prevails in HF field of the spectrum. In other words, the presence of absorption is displayed in additional spreading of LF spectrum region, and the presence of nonlinearity – in spreading of HF range.

All the mentioned allowed to obtain the formula for calculation of intensity increment on basis of taking into account nonlinear – elastic soil behavior or elastic nonlinearity (at usage of vibration source):

$$\Delta I = 3 \lg A_i f_{wai} / A_o f_{wao},$$

where $A_i f_{wai}$, $A_o f_{wao}$ is the product of spectrum amplitude on weight-average vibration frequency of investigated and etalon soils.

The formula $G = G_{max} = \rho vs^2$ characterizes soil nonlinear–elastic behavior at the absence of absorption.

If the impulsive source is used at SMZ than the formula will have the form:

$$\Delta I = 2 \lg A_i f_{wai} / A_o f_{wao.}$$

Seismic Microzonation on basis of taking into Account Soil Inelastic Properties

As soil liquefaction and uneven settlement of the constructions are observed at strong earthquakes, the most actual problem of SMZ is to assess possible soil nonelasticity adequately and physically proved at intensive seismic influences.

In order to assess directly nonelasticity of soil, the special scheme of the realization of experimental investigations with gas-dynamic impulsive source GSK-6M (with two oscillators) was used. Selected location of the longitudinal profile allowed to influence alternately by two emitters from adjoining and somewhat far radiation zones. In the spectrum of soil vibrations, caused by near emitter, the HF component, which quickly attenuates with distance, predominates. In case of influence by distant emitter to the soil surface, the LF component predominates in the spectrum of vibrations. In other words, at nonlinear-elastic deformations the main energy is concentrated in the HF range of spectrum and at nonelastic – in the LF range. The signal spectrum has the symmetrical form in the far and practically linear-elastic zone.

Elastic linear and nonlinear vibrations are characterized for the given source by the constancy of the real spectrum square, which is the index of definite source energy value, absorbed by soil (which is deformed by the source). The analysis of strong and destructive earthquake records and also the analysis of specially carried out experimental influences showed that at nonelastic phenomena spectra square of corresponding soil vibrations is not the constant value. It can decrease and the more it decreases, the less the soil solidity and the greater the influence value.

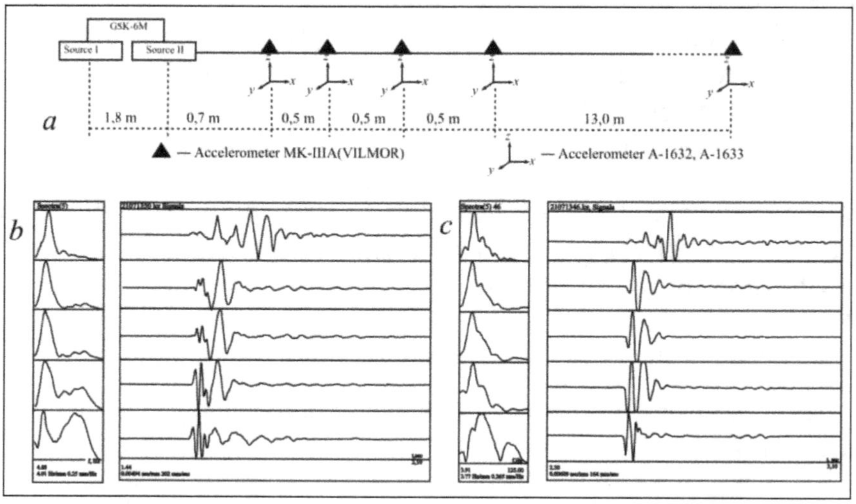

Investigation of site spectral features by means of GSK-6M seismic source:
a) Experiment scheme; b) Record of second source impact; c) Record of first source impact.

At usage of vibratory energy source, the whole number of new formulas in order to assess soil seismic hazard with taking into account the values of their nonelasticity were obtained:

$$\Delta I = 2{,}4 \lg [(S_r i)_n (S_{ro})_d / (S_r i)_d (S_{ro})_n],$$

where $(S_{ri})_{n,d}$ and $(S_{ro})_{n,d}$ are the squares of real spectra of investigated and etalon soils in near and distant zones of the source.

$$\Delta I = 3{,}3 \lg (Aif_{aw} i)_n (A_o f_{awo})_d / (Aif_{aw} i)_d (A_o f_{awo})_n,$$

where $(A_i f_{awi})_{n,d}$ and $(A_o f_{awo})_{n,d}$ are the amplitudes and weight-average frequencies of investigated and etalon soils in near and distant zones of the source.

In case of powerful impulsive source usage the offered formulas will have a form:

$$\Delta I = 1{,}2 [\lg (S_r i)_n (S_{ro})_d / (S_r i)_d (S_{ro})_n],$$

where $(S_{Pi})_{бд}$ and $(S_{Po})_{бд}$ are the squares of real spectra of investigated and etalon soils in near and distant zones of the source;

$$\Delta I = 2 \lg [(Aif_{aw} i)_n (A_o f_{awo})_d / (A\, i f_{aw} i)_d (A_o f_{awo})_n],$$

where $(A_i f_{awi})_{n,d}$ and $(A_o f_{awo})_{n,d}$ are the amplitudes and weight-average frequencies of investigated and etalon soils in near and distant zones of the source.

The formulas $\frac{G}{G_{max}} = k(\gamma, PI)(\sigma)^d$, and $\xi = 0.333 \frac{1+\exp(-0.0145PI^{1,3})}{2}\left[0.586\left(\frac{G}{G_{max}}\right)^2 - 1.547\frac{G}{G_{max}} + 1\right]$ are true only for loose dispersal soils. The formulas $\frac{G}{G_{max}} = k(\gamma, PI)(\sigma)^d$, and $\xi = 0.333 \frac{1+\exp(-0.0145PI^{1,3})}{2}\left[0.586\left(\frac{G}{G_{max}}\right)^2 - 1.547\frac{G}{G_{max}} + 1\right]$ were used at SMZ of the territory of Kutaisi city. Besides, with the help of the formulas $\Delta I = 3{,}3\lg(W_{gr} / W_{et}) +$ $3{,}3\lg(W_{top} / W_{fnd})$ and $K = -0.1 + 0.68\lg R$ nonelastic deformation properties of soils in full-scale conditions on the site of Novovoronezh NPP-2 were defined more exactly. The formulas were obtained on basis of physical principle, which underlies the scheme, applied at the assessment of soil looseness measure.

Calculational Method of Seismic Microzonation

Calculational method of SMZ is used in order to analyse features of soil behavior with introduction of definite engineering–geological structure characteristics of investigated site as initial data: values of transverse wave velocities, index of extinction, modulus of elasticity, power of soil layers, their consistency etc. Calculational method includes thin-layer medium, multiple-reflected waves, finite-difference method, finite-elements analysis (FEA) and other techniques.

One can take nonlinear soil properties into account in the problems of earthquake engineering by means of instrumental and calculation methods. The instrumental method of SMZ is the main method. Nevertheless it is quite often necessary to solve such problems using calculational method, which allows to model practically any conditions, which are observed in the nature. At the same time the practice reqirements lead to the necessity of calculation of soil vibrations for the conditions of their nonlinear-elastic and nonelastic deformations. At the solution of such problem it is assumed that elastic half-space behaves as linear-elastic medium and the covering soil displays strong nonlinear properties at intensive seismic or dynamic influences.

Instrumental stress-sstrain dependences can be used, for example one obtained for plastic clay soil shown in figure. The conception of the so-called soil bimodularity, is taken into account in the given dependence. Considerable differences in behavior of "weak" soils at compression and dilatation lie in the base of the phenomenon. Such soil is characterized at dilatation by quite small shear modulus.

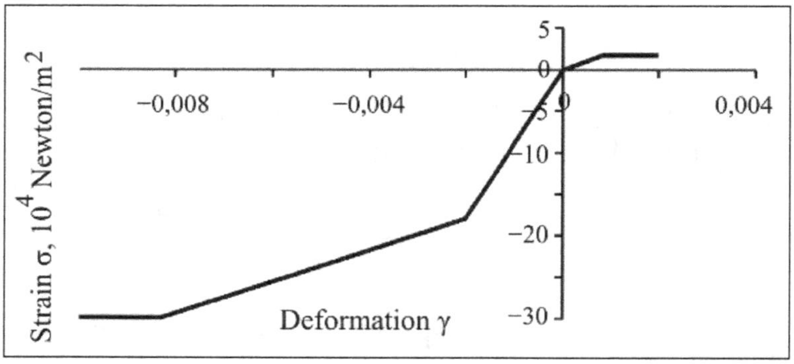

Instrumental stress-sstrain curve, showing property of soil bimodularity.

The solution of the given nonlinear problem for soils in the analytic form is based, as a rule, on considerable assumptions due to the complication of adequate taking into account behavior features of such complex system as the soil. Therefore the numerical solution of nonlinear problems on the modern stage of knowledge is the most proved if the data of field or laboratory investigations is taken into account in these or those connections. Thus, the correlations, which are determined by the experimental investigations, are the basis of the solution of calculation nonlinear problems. In other words calculation programs for the solution of calculation nonlinear problems essentially are analytical-empirical. The most adequate programs are exactly like these (SHAKE, NERA etc.)

Equivalent Linear Model: SHAKE and EERA Programs

Equivalent linear model is one of the first models, which take nonlinear soil behavior into account. Equivalent linear approximation consists in modification of the model of

Kelvin–Voight (for taking some types of nonlinearity into account) and, for example, is realized in the programs SHAKE and EERA.

Equivalent linear model is based on the hypothesis that shear modulus G and attenuation coefficient ξ are the functions of shearing strain γ. In the programs SHAKE and EERA (Equivalent-linear Earthquake site Response Analyses) the values of shear modulus G and attenuation coefficient ξ are determined (in the process of iteration) so that they correspond to the deformation levels in each layer.

IM Model: NERA Program

In 2001 realization principle, which was used in the program EERA, was applied in the programming of NERA (Nonlinear Site Response Analysis), which allows to compute soil thickness nonlinear reaction on seismic influences. The program is based on the medium model, offered by Iwan and Mroz, which is often called the IM model for short. As it is shown in the figure, the model supposes the simulating of nonlinear curves strain-deformation, using a number of n mechanical elements, which have different stiffness kj and sliding resistance R_j, where $R_1 < R_2 < ... < R_n$. Initially the residual stresses in all elements are equal to zero. At monotonically increasing load the element j deforms until the transverse strain τ reaches R_j. After that the element j keeps positive residual stress, which is equal to R_j. The equation, describes dynamics of soil medium, is solved by the method of central differences.

Calculation of Nonlinear Absorptive Ground Medium Vibrations

Let's suppose that we have the seismic wave, which falls on the soil thickness surface. Let's assume that soil thickness is nonlinear absorptive unbounded medium with the density (and S-wave propagation velocity v_S. At small deformations the value of shear modulus G will be maximum for the given soils:

$$G = G_{max} = \rho vs^2$$

At the deformation increase the value G remains constant at first but at reaching some value (which is definite for each material or soil) the value G considerably changes, i.e. the soil begins to display its nonlinear properties. At the continued deformation increase the growth of stresses decelerates and then can remain unchanged until material destruction or hardening, i.e. until structural condition change.

As the main soil index, which characterizes its type and behavior at intensive loads, the value of plasticity PI was chosen. The parameters, which are necessary for calculations, are determined on basis of empirical ratios:

$$k\ \gamma,PI\ = 0.5\left\{1 + \tanh\left[\ln\frac{0.000102\quad n\ PI}{}\right]0.492\right\}$$

$$\text{where } n(PI)=\begin{cases}0.0 & \text{for } PI=0,\\ 3.37\cdot 10^{-6}PI^{1.404} & \text{for } 0<PI\le 15,\\ 7.0\cdot 10^{-7}PI^{1.976} & \text{for } 15<PI\le 70,\\ 2.7\cdot 10^{-5}PI^{1.115} & \text{for } PI>70\ ;\end{cases}$$

$$d=0.272\left\{1-\tanh\left[\ln\left(\frac{0.000556}{\gamma}\right)^{0.4}\right]\right\}e^{-0.0145PI^{1.3}}$$

Then the change of shear modulus is determined on basis of the ratio,

$$\frac{G}{G_{max}}=k(\gamma,PI)(\sigma)^{d},$$

where G is the current shear modulus, is normal stress.

Seismic energy absorption is calculated by the formula,

$$\xi=0.333\frac{1+\exp(-0.0145PI^{1,3})}{2}\left[0.586\left(\frac{G}{G_{max}}\right)^{2}-1.547\frac{G}{G_{max}}+1\right]$$

On basis of the given ratios and introduced by us ratios for determination of necessary indices (normal stress, deformation etc), nonlinear version of the program ZOND was worked out. From the database of strong motions AGESAS, which was formed by us, the accelerogram, which was recorded on rocks in Japan, with the characteristics (magnitude, epicentral distance, spectral features etc.) similar to the territory of Tbilisi city, was chosen as the accelerogram, given into the bedrock.

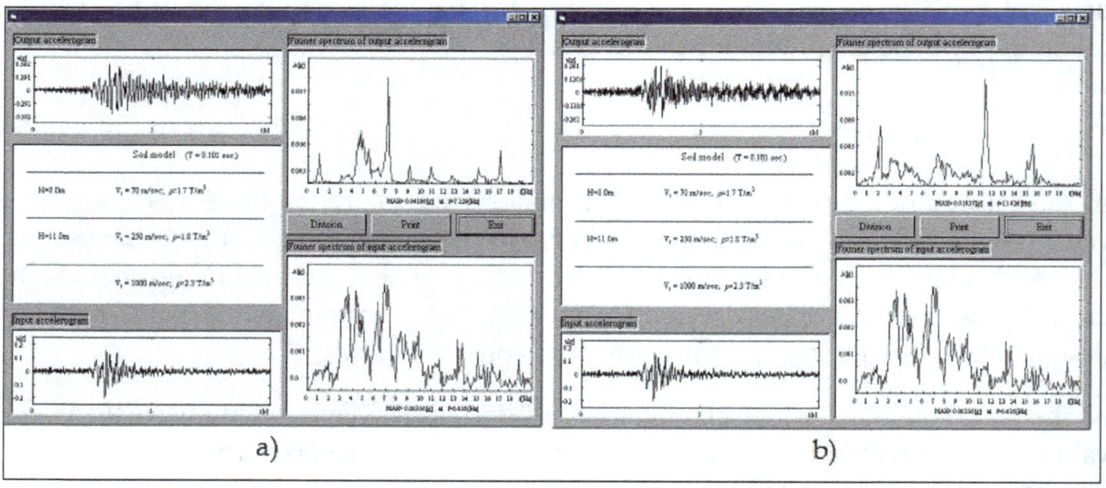

Results of calculations using multiple reflected waves' tool in linear (a) and nonlinear (b) cases.

The analysis of the results of linear and nonlinear calculations models of definite areas of Tbilisi city territory confirms the adequacy of calculations to the physical phenomena, which were obtained in soils at intensive loads. With the increase of seismic influence intensity the nonlinearity display increases. Absorption grows simultaneously. Hence the resulting motion at quite high influence levels can be lower than the initial level. It corresponds to the fact, which is known on the results of analysis of strong earthquake consequences, which happened in recent years.

Calculation of Nonlinear Soil Response using FEM Tool of Seismic Microzonation

The problem of the determination of soil massif response on dynamic influence with taking soil nonlinear properties into account can be solved by usage of finite element method (FEM) in the following way.

Soil medium is represented in the form of two-dimensional massif, which is approximate by triangular finite elements. The net, which consists of triangular elements, allows to describe quite accurately any relief form and form of the layer structure of soil massif with its physics-mechanical parameters. Within finite elemet the soil is homogeneous with inherent to it characteristics, which vary in time depending on influence intensity. Earthquake accelerogram of horizontal or vertical direction, which is applied, as a rule, to the foundation of soil massif, is used as the influence. Soil is in the conditions of plane deformation and is considered as an orthotropic medium. Axes of the orthotropy coincide with the directions of main strains.

The problem of nonlinear dynamics of soil massif is solved by means of the consecutive determination of mode of deflection of the system on the previous step. The system is linear-elastic on each step.

Instrumental-calculational Method of Seismic Microzonation

In recent years a new "instrumental-calculational" method of SMZ (per se simultaneously having the features of both instrumental and calculational method) which includes tool of "instrumental-calculation analogies" has been developed in Russia in recent years. Its usage is based on direct usage of modern databases of strong motions.

As a basis at realization of tool instrumental database of strong movements, registered in definite soil conditions, is used. As a result of given database with the help of numerical calculations it is possible more or less safety to forecast behavior of these or those soils (or their combination) for strong (weak) earthquakes with typical characteristics for the investigated territory (magnitude, epicentral distance, focus depth etc.).

Relief Influence on the Earthquake Intensity in SMZ Problems

Morphological and morphometric features of relief meso- and macroforms influences on seismic intensity increment.

On basis of the analysis of numerous macroseismic observations the consequences of strong earthquakes, which took place on the territory of the former USSR, S.V.Puchkov and D.V.Garagozov offered the empirical formula for the intensity increment calculation (ΔI) depending on relief feature:

$$\Delta I = 3{,}3\lg(W_{gr} / W_{et}) + 3{,}3\lg(W_{top} / W_{fnd}),$$

where W_{gr}, W_{et} are the accelerations of vibratory motion on soil and etalon; W_{top}, W_{fnd} are the accelerations on the top of mountain construction and its foundation.

It was determined as a result of the instrumental and theoretical investigations that for the microrelief the increment of seismic intensity increases from the foundation of mountain-shaped feature to its top and can reach approximately 1.8 degree. For the locality mesorelilef the tendency of the increase of seismic vibration intensity from foundation to the top remains. The increment of seismic intensity for the relief mesoforms is about 0.3 degree. It was shown that weak hilly relief, with the inclinations less than 10°, does not influence on the seismic vibrations intensity.

The investigations of S.V.Puchkov and D.V.Garagozov showed that at vibrations of mountain range, composed by volcanic tuf, the amplitude of seismic vibrations in S-waves increases on the height 15 m in 1.46 times in comparison with the foundation. For the massif, composed by loamy sand and loams on the same height marks the vibrational amplitude increased in 1.8 times for p-waves and in 3.2 times for S-waves.

Slope steepness considerably influences on the increment of seismic intensity. The increase of slope steepness, composed by incoherent gravel-pebble and sabulous-loamy grounds is conductive to the sharp worsening of engineering-geological and seismic conditions of the territory. So, for example, it is determined that slope steepness more than 19°–15° (for dry sandy-argillaceous and gravel-pebble differences) produces the intensity increment up to 1 degree and at variation of slope steepness from 10° to 40° the amplitudes of seismic vibrations increase approximately in 2.5 times.

It is known that the increase of slope steepness from 40° to 80° produces the increment of seismic intensity equal to 1.5 degree.

The correlation analysis of the dependence of seismic intensity increment on true altitude, slope steepness and relief roughness showed that the main factors, which change the value of seismic intensity, are the first two indices. It conforms well to the investigation results of V.B.Zaalishvili, who introduced the new parameter of the relief coefficient.

Later the data analysis allowed to offer us the empirical formula for the possible amplification calculation K and intensity increment ΔI, which are caused by the relief:

$$K = -0.1 + 0.68\lg R$$

where R= H is the relief coefficient; is the relief slope angle, degree; H is height, m.

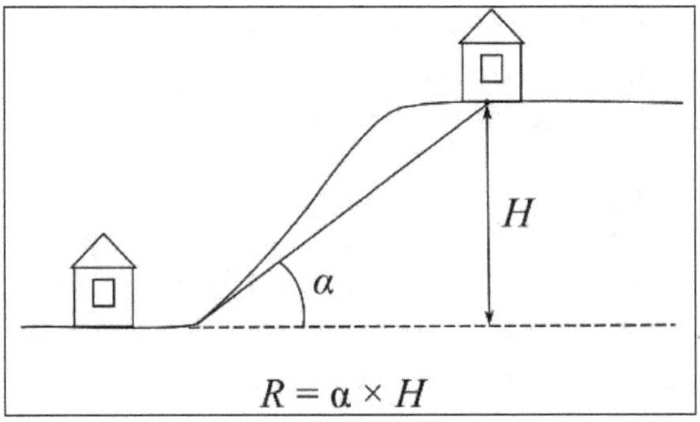

Relief coefficient R.

The analysis of the experimental data shows that intensity increment can vary at that independently of the type of rocks, from 0 to 1.5 degree. Finally, let's try to assess the amplification of vibrational amplitude, which is caused by relief, with the help of the calculation method of FEM.

The algorithm for the calculation of seismic reaction of soil thickness for the two-dimensional model was developed for this purpose. The results of the executed earliear investigations were used for the program testing. Mountain structure had the form of frustum of a cone with the height 30 m and slope angle of the generatrix 30°. The element maximum size was equal to 5 m, S-wave propagation velocity was 300 m/s, the density 1800 kg/m³. The seismic influence was applied to the foundation of soil thickness in the form of instrumental accelerogram, modeling the vertically propagating SH wave.

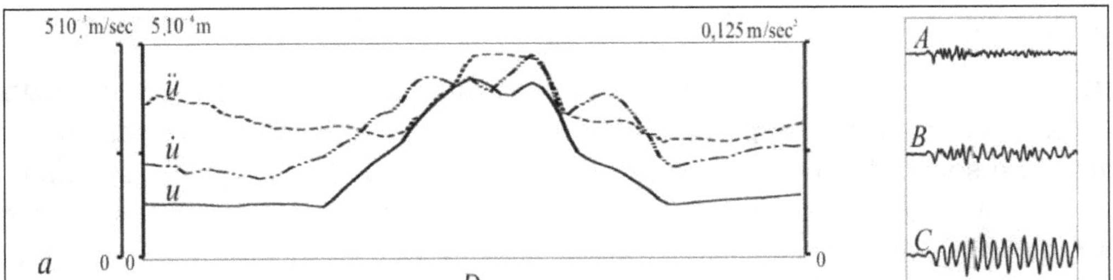

Final elements analysis (FEA) application example: a) Variation of amplitudes of displacement, velocity and acceleration along surface; b) calculational model; c) seismograms, calculated in points A, B, C, D.

It was determined that the vibrational amplitude considerably chances with the relief. The given dependence at that is various for the displacements, velocities and accelerations. The largest value of the amplification is observed for displacements and the maximum ratio of vibrational amplitudes, for example, in the point C to the point A, is 2.1 and for the point D – 3.2. It well satisfies the results of experimental observations where the ratio in the point C for the S-wave is equal to 2.3 and in the spectral region the maximum values are 1.8 (at T = 0.4 s) and 3.2 (at T = 0.7 s) for P- and S-waves accordingly. Spectral analysis also shows the resonance increase of vibrational amplitudes in the top part of the slope on the frequency 1.6 Hz (i.e. T=0.6 s).

Considerably fewer investigations are dedicated to the influence of the underground relief on the intensity. On the data of B.A.Trifonov the underground and buried topography of the rocks influences on seismic vibrations intensity, if the surface slope exceeds 0.3. At the vee couch of the rocks, which are covered by sedimentary thickness, the ratio between wave length and the sizes of vee stripping influences on seismic intensity change. Seismic intensity increment in the given case is formed by the wave interference and can be 1.5–2.0 degree.

Thus, at the execution of SMZ works in the mountain regions or under the conditions of billowy relief, it is necessary to pay special attention to the influence of surface or underground relief on the intensity forming. It is necessary to continue the investigations in order to obtain statistically proved ratio for the calculation of intensity increment, caused by relief.

Seismic Microzonation of Vladikavkaz City

If we consider 5% DSZ map as basis for seismic microzonation so seismic intensity of 8 corresponds to etalon grounds for whole territory. Then, maps of seismic microzonation of cities must be created. According to the above mentioned maps of detailed zoning the maps of seismic microzonation with probability 1%, 2%, 5% or 10 %, correspondingly, were made up.

Though, that definitions of the word "zoning" are similar, actually they are quite different in essence. Unlike the maps of detailed seismic zoning, which give seismic potential (Mmax) and source features, the maps of seismic microzonation give assessments of soil condition influence (sands, rocks, pebbles, clays etc., their combination; watering; relief (as underground as surface); spectral distribution of incoming wave; predominant vibration frequencies on city square etc.) on forming of future earthquake intensity. As a rule, the scale of such maps is 1:10 000, in order to have the opportunity of taking them into account at building. Maps can be more detailed (1:5000 etc.) but this makes no sense as the type and physical condition of soils in space on the territory site can change fast. The most important thing is to assess intensity of possible earthquakes on areas with typical soil conditions for city territory.

Maps of seismic microzonation can be made up for the certain territories (cities and settlements, as a rule). It is impossible to make them up in entire format because of the necessity of geological conditions knowledge on larger territories, which are mostly not built up. We often don't have such data even for the modern cities! It's practically impossible because the resources will be lost for nothing! And absurdity! In the other words there is no the microzonation map even for the territory of North Ossetia let alone the whole Northern Caucasus.

Maps of seismic microzonation do not only show the place of earthquake-proof building up, but they also show on what intensity this or that building must be calculated and designed: on 6, 7, 8 or 9 points. And sometimes even on 10 points (for very soft grounds!). And this suggests investments of different financing for the realization of antiseismic measures (thicker armature, more connections etc.). Seismic risk can considerably be reduced at building-up zones with 7, 8 and 9 point of the calculated intensity by adequate site development on the territory of city, for example, as social losses will be minimal, though buildings will be damaged in this or that extent.

In the next stage we should carry out SMZ. It should be noted that as a basis the maps of different probability of exceedance will be used and as the initial intensity, the value of which corresponds directly to the intensity of the sites, composed by average soils or characterized by average soil conditions and, therefore, the maps will be referred to the 7, 8 or 9 points (and similarly for acceleration). The zones, composed by clay soils of fluid consistency, which can be characterized by liquefaction at quite strong influences, are marked by the index 9*. Intensity calculation here supposes the usage of special approaches in the form of direct taking soil nonlinearity into account. The usage of relevant methods and techniques of SMZ will allow to obtain the correspondent maps of SMZ.

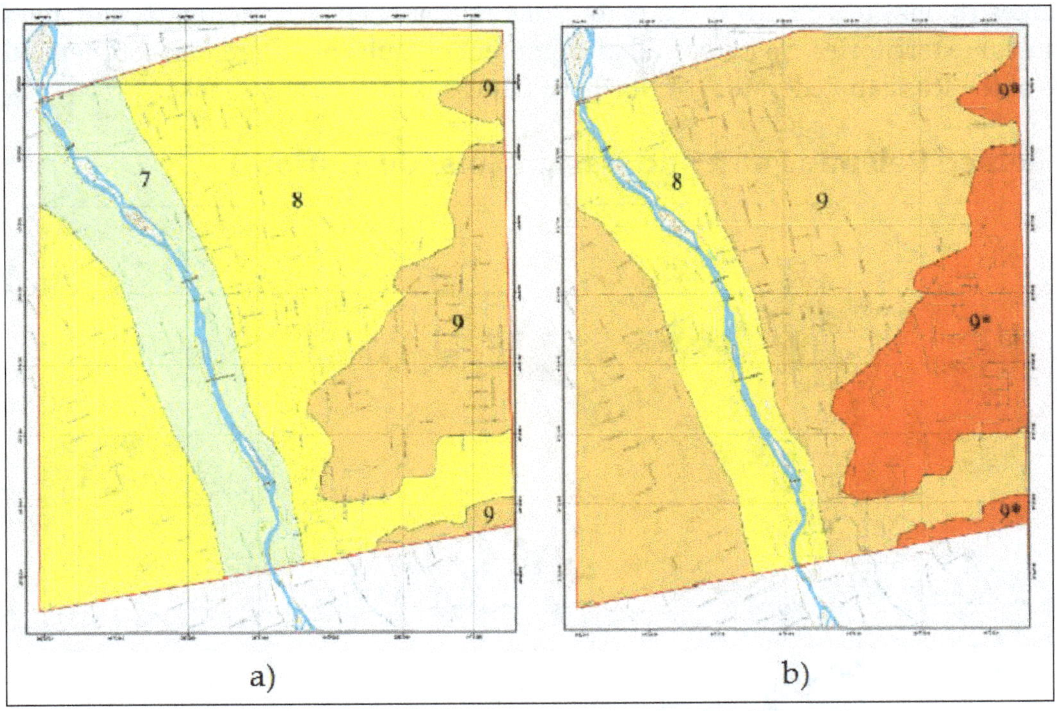

The maps of seismic intensity microzonation for probabilities of 5% (a) and 2% (b) for the central part of Vladikavkaz city territory.

Thus for maps with probability of exceedance 1%, 2%, 5% and 10% one can obtain corresponding maps of SMZ with probability of exceedance 1%, 2%, 5% and 10%, i.e. probabilistic maps of SMZ.

For each of the zoning subject the probabilistic map of the seismic microzonation with location of different calculated intensity (7, 8, 9, 9*) zones is developed (the zones, composed by clay soils of fluid consistency, which can be characterized by liquefaction at quite strong influences, are marked by the index 9*). The maps in accelerations units show the similar results.

Such maps of SMZ except of mentioned developments are also based on materials of local network of seismic observations "Vladikavkaz". Network was organized for the first time on the urbanized territory of the Northern Caucasus in July 2004. Stations are located on the sites with different typical for the city soils (clays of medium-hard and liquid consistence, gravels with filling material of less than 30% and more than 30%, and their assembly).

It must be noted that usage of the maps with high time exposition i.e. maximal magnitude (maximal intensity) for given territory (for return period of 50 years and exceedance probability 2% or 1%) physical nonlinearity of soils necessarily must be taken into account with the help of developed tools.

Unlike small-scale M 1:8 000 000 seismic hazard map of the territory of Russia (GSZ) maps of DSZ in scale 1:200 000 allow taking into account features of specific seismic sources (faults) directly. But the main thing is that such scale zoning is suitable for quite large territories. So it's seen that alignment of faults of different constituent entities of the Russian Federation of Northern Caucasus make a good sense.

Specified Seismic Fault and Design Seismic Motion

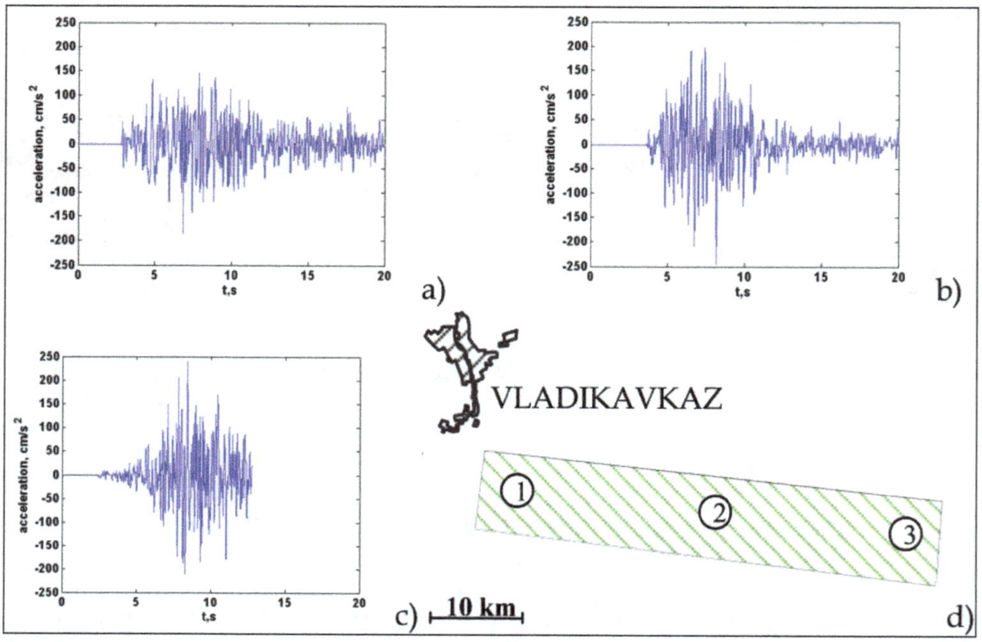

Synthetical accelerograms for different source locations: a – western part of fault; Sb – middle part of fault; c – eastern part of fault; d – scheme of sources of scenarios earthquakes.

Analysis and consequent account of initial accelerograms transformation will become the basis for site effect analysis at strong seismic loadings. Methods of such modeling are based on accordance of spectral properties of modeled and real earthquake. In a whole modeling accuracy depending on the purposes of total motion usage and what characteristics defining structural system behavior must be reproduced.

Earthquake source that is a region of rupture can be considered as point source only for much larger distances than fault size. At close distances effects of finite fault size become more significant. Those phenomena are mainly connected with finite rupture velocity, which causes energy radiation of different fault parts in different times and seismic waves are interference and causes directivity effects.

Let's compare amplitude spectra of obtained design accelerograms with spectrum of real earthquake from considered fault. Data analysis shows that spectra of calculated and real earthquakes in a whole are similar in their main parameters.

It must be noted that spectrum of vertical component of real earthquake is closer to design spectra. The last fact is quite obvious and is explained by proximity to earthquake source. Indeed, close earthquakes in general are characterized by predomination of vertical component. Record of TEA station (located in theater) was selected due to its location on dense gravel and has a minimal distortions caused by soil conditions.

Analysis of spectrum of weak earthquake shows that peaks are observed on 1.3 and 5.6 Hz. In spectra of synthesize accelerograms mentioned amplitudes are also observed. At the same time medium response on strong earthquake, undoubtedly, differ from weak earthquake response.

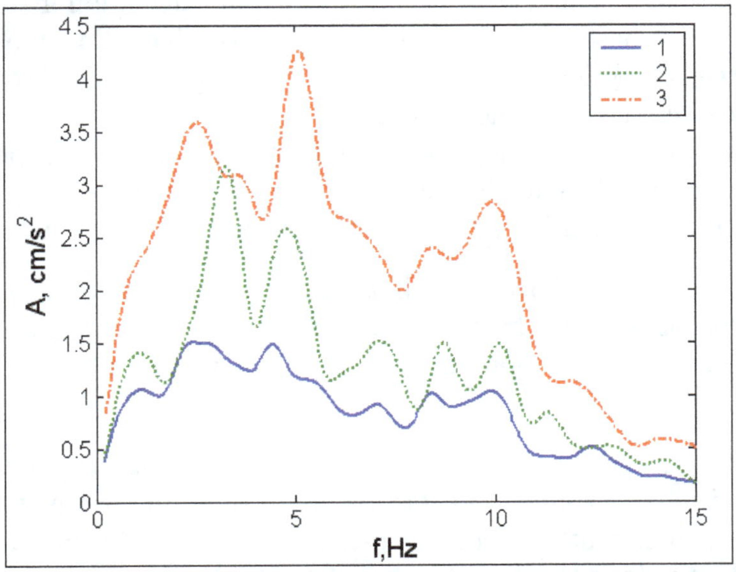

Spectra of design accelerograms at different source locations of earthquake M=7,1: 1 – western part of fault; 2 – middle part of fault; 3 – eastern part of fault.

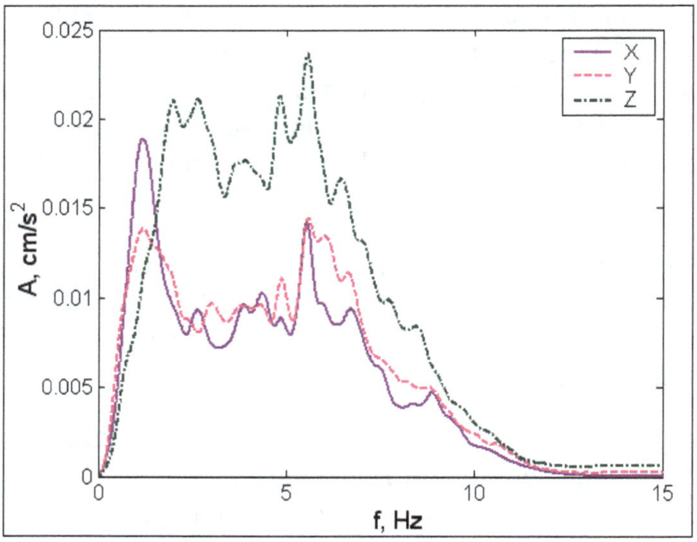

Spectra of accelerograms of weak earthquake with epicenter in the zone of Vladikavkaz fault. (25.08.2005 10:25 GMT, H = 8 km M= 2.5).

Usage of maps of detailed seismic zoning in units of accelerations at seismic microzonation level is possible only for calculation method giving results in units of accelerations. Today traditional instrumental method of seismic microzonation does not allow obtaining intensity increments in accelerations due to traditional orientation on macroseismic intensity indexes. The exclusion is the case of investigation of strong earthquakes accelerations when instrumental records are obtained (in presence of accelerometer). At the same time investigations are conducted and the problem supposed to be solved.

On the other hand in recent years a new instrumental-calculation method was developed. New method is based on selection from database (including about 5000 earthquake records) soil conditions which are the most appropriate to real soil conditions of the investigated site. Then the selection of seismic records with certain parameters or their intervals follows (magnitude, epicentral distance, and source depth). Then maximal amplitudes are recalculated for given epicentral distances. Absorption coefficient can be calculated by attenuation model for given region.

Thus, a new complex method of seismic hazard assessment providing probability maps of seismic microzonation, which are the basis of earthquake-proof construction, is introduced. Undoubtedly such approach significantly increases physical validity of final results.

Considered procedures on the level of possible seismic sources zones exploration, maps of detailed seismic zoning and seismic microzonation may differ. So paleoseismological investigations like trenching, which allow determining more reasonable the recurrence and other features of seismic events realization are also possible when it is necessary.

Today, we have conditions for detailed seismic zoning maps development like the but for all the territory of the Northern Caucasus on basis of the modern achievements of engineering seismology. It will give us a possibility to develop probabilistic maps of seismic microzonation with the help of powerful nonexplosive sources, methods taking into account physical soils nonlinearity.

Thus algorithm of seismic hazard assessment of the territory taking into account multiple factors forming seismic intensity was considered. Forms of typical seismic loadings for firm soils are given, which will be changed from site to site in dependence of differences in ground conditions (engineering-geological, geomorphological and gidrogeological conditions).

References

- Principles-of-probabilistic-seismic-hazard-assessment-psha-and-site-effect-evaluation-and-its-applic, earthquakes-forecast-prognosis-and-earthquake-resistant-construction: intechopen.com, Retrieved 22 April, 2020
- Musson R. M. W. and Henni P. H. O., Methodological considerations of probabilistic seismic hazard mapping, Soil Dynamics and Earthquake Engineering 21, 385–403, 2001
- NEELIMA SATYAM. D and K. S. RAO, "Estimation of Peak Ground Acceleration for Delhi Region using Finsim, a Finite Fault Simulation Technique" International Journal of Earth Sciences and Enginee ring ISSN 0974-5904, Vol. 02, No. 03, pp. 215-223, July 2009
- Assessment-of-seismic-hazard-of-territory, earthquake-engineering: intechopen.com, Retrieved 8 March, 2020

5

Seismic Protection

Seismic protection is referred to as the sophisticated design approach involved in the construction to minimize the damages to structure during earthquake. It includes seismic base isolation, seismic loading, seismic retrofit, seismic metamaterial, viscoelastic and fluid viscous damper. This chapter closely examines seismic protection to provide an extensive understanding of the subject.

The seismic protection of the structures is one of the most attractive goals of the structural engineers in order to save lives and minimize damages to structures in case of earthquakes of high intensity.

The absence of damage is of primary importance for some types of structures, such as:

- Structures which house high risk components and materials (nuclear facilities, some chemical plants, etc).
- Important bridges and viaducts.
- Important public buildings (hospitals, emergency control centres, energy and communication distribution centres, etc).
- Structures which are important for the national defence.
- Structures which house costly components and electronic equipment.
- Museums and historic and artistic monuments.

Isolators

They have 3 main purposes: bear the weight of the structures, reduce the period of the structures and consequently the seismic action, dissipate energy:

- HDRB: High Damping Rubber Bearings.
- LRB: Lead Rubber Bearings.
- PS: Sliding pendulum system.
- PND/PNUD: Hysteretic Isolators with pot bearings.

- CSD/CSUD: Hysteretic isolators with spherical bearing.

Dampers

Their main purpose is to dissipate seismic energy, in conjunction with or even without the use of isolators:

- Hysteretic Dampers.
- Elastic Dampers.
- Viscous Dampers.
- Visco-Elastic Dampers.
- STU - Shock Transmission Units.
- Hysteretic bracing systems.

Base Isolation for the Buildings

Base isolation is becoming more and more a widespread technique to protect structures from the devastating effects of earthquakes.

One of the leading causes of death of people during an earthquake is the collapse of houses and other buildings. The design of a building carried out by the latest earthquake standards, leads to design the structural elements so that they can resist earthquake without collapsing, even if few damages can occur.

This approach, however, does not care about the non-structural elements (floors, curtain walls, windows) or the content of the structure (equipment, furnishings, furniture). What is then to have a hospital after an earthquake still standing but with the high-tech equipment destroyed, floors and curtain walls cracked? The costs of stopping the activity (just when the seismic event would require) and repair after an earthquake are considerable. The seismic isolation technique isolates the structure from the earthquake preventing its devastating energy to penetrate inside.

In this way the structure can be lightened, with considerable savings and, in case of earthquake, all content and non-structural parts are safeguarded. The damages after the earthquake will be so non-existent or marginal and the structure functionality is guaranteed.

SEISMIC ANALYSIS

Seismic analysis is a subset of structural analysis and is the calculation of the response of a building (or nonbuilding) structure to earthquakes. It is part of the process of structural design, earthquake engineering or structural assessment and retrofit in regions where earthquakes are prevalent.

A building has the potential to 'wave' back and forth during an earthquake (or even a severe wind storm). This is called the 'fundamental mode', and is the lowest frequency of building response. Most buildings, however, have higher modes of response, which are uniquely activated during earthquakes. The figure just shows the second mode, but there are higher 'shimmy' (abnormal vibration) modes. Nevertheless, the first and second modes tend to cause the most damage in most cases.

The earliest provisions for seismic resistance were the requirement to design for a lateral force equal to a proportion of the building weight (applied at each floor level). This approach was adopted in the appendix of the 1927 Uniform Building Code (UBC), which was used on the west coast of the United States. It later became clear that the dynamic properties of the structure affected the loads generated during an earthquake. In the Los Angeles County Building Code of 1943 a provision to vary the load based on the number of floor levels was adopted (based on research carried out at Caltech in collaboration with Stanford University and the U.S. Coast and Geodetic Survey, which started in 1937). The concept of "response spectra" was developed in the 1930s, but it wasn't until 1952 that a joint committee of the San Francisco Section of the ASCE and the Structural Engineers Association of Northern California (SEAONC) proposed using the building period (the inverse of the frequency) to determine lateral forces.

Earthquake engineering has developed a lot since the early days, and some of the more complex designs now use special earthquake protective elements either just in the foundation (base isolation) or distributed throughout the structure. Analyzing these types of structures requires specialized explicit finite element computer code, which divides time into very small slices and models the actual physics, much like common video games often have "physics engines". Very large and complex buildings can be modeled in this way (such as the Osaka International Convention Center).

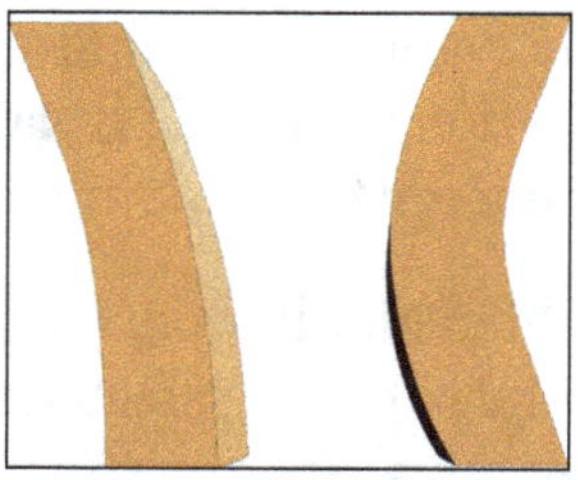

First and second modes of building seismic response.

Structural analysis methods can be divided into the following five categories:

Equivalent Static Analysis

This approach defines a series of forces acting on a building to represent the effect of earthquake ground motion, typically defined by a seismic design response spectrum. It assumes that the building responds in its fundamental mode. For this to be true, the building must be low-rise and must not twist significantly when the ground moves. The response is read from a design response spectrum, given the natural frequency of the building (either calculated or defined by the building code). The applicability of this method is extended in many building codes by applying factors to account for higher buildings with some higher modes, and for low levels of twisting. To account for effects due to "yielding" of the structure, many codes apply modification factors that reduce the design forces (e.g. force reduction factors).

Response Spectrum Analysis

This approach permits the multiple modes of response of a building to be taken into account (in the frequency domain). This is required in many building codes for all except very simple or very complex structures. The response of a structure can be defined as a combination of many special shapes (modes) that in a vibrating string correspond to the "harmonics". Computer analysis can be used to determine these modes for a structure. For each mode, a response is read from the design spectrum, based on the modal frequency and the modal mass, and they are then combined to provide an estimate of the total response of the structure. In this we have to calculate the magnitude of forces in all directions i.e. X, Y & Z and then see the effects on the building.. Combination methods include the following:

- Absolute: Peak values are added together.
- Square root of the sum of the squares (SRSS).
- Complete quadratic combination (CQC): A method that is an improvement on SRSS for closely spaced modes.

The result of a response spectrum analysis using the response spectrum from a ground motion is typically different from that which would be calculated directly from a linear dynamic analysis using that ground motion directly, since phase information is lost in the process of generating the response spectrum.

In cases where structures are either too irregular, too tall or of significance to a community in disaster response, the response spectrum approach is no longer appropriate, and more complex analysis is often required, such as non-linear static analysis or dynamic analysis.

Linear Dynamic Analysis

Static procedures are appropriate when higher mode effects are not significant. This is generally true for short, regular buildings. Therefore, for tall buildings, buildings with torsional irregularities, or non-orthogonal systems, a dynamic procedure is required. In the linear dynamic procedure, the building is modelled as a multi-degree-of-freedom

(MDOF) system with a linear elastic stiffness matrix and an equivalent viscous damping matrix.

The seismic input is modelled using either modal spectral analysis or time history analysis but in both cases, the corresponding internal forces and displacements are determined using linear elastic analysis. The advantage of these linear dynamic procedures with respect to linear static procedures is that higher modes can be considered. However, they are based on linear elastic response and hence the applicability decreases with increasing nonlinear behaviour, which is approximated by global force reduction factors.

In linear dynamic analysis, the response of the structure to ground motion is calculated in the time domain, and all phase information is therefore maintained. Only linear properties are assumed. The analytical method can use modal decomposition as a means of reducing the degrees of freedom in the analysis.

Nonlinear Static Analysis

In general, linear procedures are applicable when the structure is expected to remain nearly elastic for the level of ground motion or when the design results in nearly uniform distribution of nonlinear response throughout the structure. As the performance objective of the structure implies greater inelastic demands, the uncertainty with linear procedures increases to a point that requires a high level of conservatism in demand assumptions and acceptability criteria to avoid unintended performance. Therefore, procedures incorporating inelastic analysis can reduce the uncertainty and conservatism.

This approach is also known as "pushover" analysis. A pattern of forces is applied to a structural model that includes non-linear properties (such as steel yield), and the total force is plotted against a reference displacement to define a capacity curve. This can then be combined with a demand curve (typically in the form of an acceleration-displacement response spectrum (ADRS)). This essentially reduces the problem to a single degree of freedom (SDOF) system.

Nonlinear static procedures use equivalent SDOF structural models and represent seismic ground motion with response spectra. Story drifts and component actions are related subsequently to the global demand parameter by the pushover or capacity curves that are the basis of the non-linear static procedures.

Nonlinear Dynamic Analysis

Nonlinear dynamic analysis utilizes the combination of ground motion records with a detailed structural model, therefore is capable of producing results with relatively low uncertainty. In nonlinear dynamic analyses, the detailed structural model subjected to a ground-motion record produces estimates of component deformations for each degree of freedom in the model and the modal responses are combined using schemes such as the square-root-sum-of-squares.

In non-linear dynamic analysis, the non-linear properties of the structure are considered as part of a time domain analysis. This approach is the most rigorous, and is required by some building codes for buildings of unusual configuration or of special importance. However, the calculated response can be very sensitive to the characteristics of the individual ground motion used as seismic input; therefore, several analyses are required using different ground motion records to achieve a reliable estimation of the probabilistic distribution of structural response. Since the properties of the seismic response depend on the intensity, or severity, of the seismic shaking, a comprehensive assessment calls for numerous nonlinear dynamic analyses at various levels of intensity to represent different possible earthquake scenarios. This has led to the emergence of methods like the incremental dynamic analysis.

SEISMIC BASE ISOLATION

Seismic base isolation, also known as base isolation or base isolation system, is one of the most popular means of protecting a structure against earthquake forces. It is a collection of structural elements which should substantially decouple a superstructure from its substructure resting on a shaking ground thus protecting a building or non-building structure's integrity.

Base isolation is one of the most powerful tools of earthquake engineering pertaining to the passive structural vibration control technologies. The isolation can be obtained by the use of various techniques like rubber bearings, friction bearings, ball bearings, spring systems and other means. It is meant to enable a building or non-building structure to survive a potentially devastating seismic impact through a proper initial design or subsequent modifications. In some cases, application of base isolation can raise both a structure's seismic performance and its seismic sustainability considerably. Contrary to popular belief base isolation does not make a building earthquake proof.

Base isolation system consists of *isolation units* with or without *isolation components*, where:

- Isolation units are the basic elements of a base isolation system which are intended to provide the aforementioned decoupling effect to a building or non-building structure.
- Isolation components are the connections between isolation units and their parts having no decoupling effect of their own.

Isolation units could consist of shear or sliding units. This technology can be used for both new structural design and seismic retrofit. In process of seismic retrofit, some of the most prominent U.S. monuments, e.g. Pasadena City Hall, San Francisco City Hall, Salt Lake City and County Building or LA City Hall were mounted on *base isolation systems*. It required creating rigidity diaphragms and moats around the buildings, as well as making provisions against overturning and P-Delta Effect.

Base isolation is also used on a smaller scale—sometimes down to a single room in a building. Isolated raised-floor systems are used to safeguard essential equipment against earthquakes. The technique has been incorporated to protect statues and other works of art.

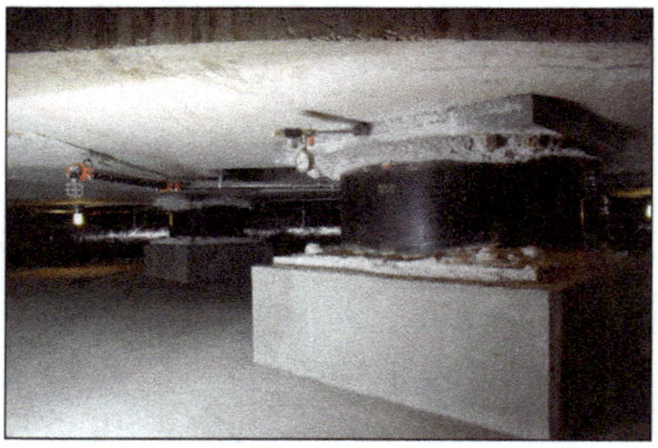

The base isolators under the Utah State Capitol building.

LA City Hall, the world's tallest building to be retrofitted with base isolation.

Through the George E. Brown, Jr. Network for Earthquake Engineering Simulation (NEES), researchers studied the performance of base isolation systems. The project, a collaboration among researchers at University of Nevada, Reno; University of California, Berkeley; University of Wisconsin, Green Bay; and the University at Buffalo is conducted a strategic assessment of the economic, technical, and procedural barriers to the widespread adoption of seismic isolation in the United States. NEES resources have been used for experimental and numerical simulation, data mining, networking and collaboration to understand the complex interrelationship among the factors controlling the overall performance of an isolated structural system. This project involved earthquake shaking table and hybrid tests at the NEES experimental facilities at the University of California, Berkeley, and the University at Buffalo, aimed at understanding ultimate performance limits to examine the propagation of local isolation failures (e.g., bumping against stops, bearing failures, uplift) to the system level response. These tests included a full-scale, three-dimensional test of an isolated 5-story steel building on the E-Defense shake table in Miki, Hyogo, Japan.

Adaptive Base Isolation

An adaptive base isolation system includes a tunable isolator that can adjust its properties based on the input to minimize the transferred vibration. Magnetorheological fluid dampers and isolators with Magnetorheological elastomer have been suggested as adaptive base isolators.

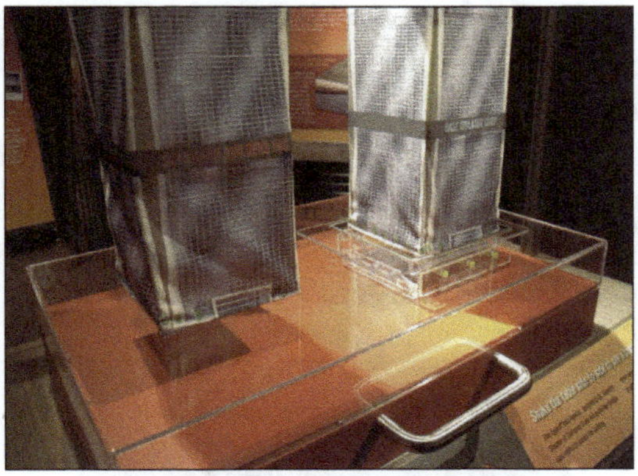

Base isolation demonstration.

SEISMIC LOADING

Seismic loading is one of the basic concepts of earthquake engineering which means application of an earthquake-generated agitation to a structure. It happens at contact surfaces of a structure either with the ground, or with adjacent structures, or with gravity waves from tsunami.

Seismic loading depends, primarily, on:

- Anticipated earthquake's parameters at the site - known as seismic hazard.
- Geotechnical parameters of the site.
- Structure's parameters.
- Characteristics of the anticipated gravity waves from tsunami (if applicable).

Sometimes, seismic load exceeds ability of a structure to resist it without being broken, partially or completely Due to their mutual interaction, seismic loading and seismic performance of a structure are intimately related.

SEISMIC RETROFIT

Seismic retrofitting is the modification of existing structures to make them more resistant to seismic activity, ground motion, or soil failure due to earthquakes. With better understanding of seismic demand on structures and with our recent experiences with large earthquakes near urban centers, the need of seismic retrofitting is well acknowledged. Prior to the introduction of modern seismic codes in the late 1960s for developed countries (US, Japan etc.) and late 1970s for many other parts of the world

(Turkey, China etc.), many structures were designed without adequate detailing and reinforcement for seismic protection. In view of the imminent problem, various research work has been carried out. State-of-the-art technical guidelines for seismic assessment, retrofit and rehabilitation have been published around the world – such as the ASCE-SEI 41 and the New Zealand Society for Earthquake Engineering (NZSEE)'s guidelines. These codes must be regularly updated; the 1994 Northridge earthquake brought to light the brittleness of welded steel frames, for example.

External bracing of an existing reinforced concrete parking garage (Berkeley).

The retrofit techniques outlined here are also applicable for other natural hazards such as tropical cyclones, tornadoes, and severe winds from thunderstorms. Whilst current practice of seismic retrofitting is predominantly concerned with structural improvements to reduce the seismic hazard of using the structures, it is similarly essential to reduce the hazards and losses from non-structural elements. It is also important to keep in mind that there is no such thing as an earthquake-proof structure, although seismic performance can be greatly enhanced through proper initial design or subsequent modifications.

Port Authority Bus Terminal.

Strategies

Seismic retrofit (or rehabilitation) strategies have been developed in the past few decades following the introduction of new seismic provisions and the availability of advanced materials (e.g. fiber-reinforced polymers (FRP), fiber reinforced concrete and high strength steel):

- Increasing the local capacity of structural elements: This strategy recognises the inherent capacity within the existing structures, and therefore adopt a more cost-effective approach to selectively upgrade local capacity (deformation/ductility, strength or stiffness) of individual structural components.
- Selective weakening retrofit: This is a counter-intuitive strategy to change the inelastic mechanism of the structure, while recognising the inherent capacity of the structure.
- Allowing sliding connections such as passageway bridges to accommodate additional movement between seismically independent structures.
- Addition of seismic friction dampers to simultaneously add damping and a selectable amount of additional stiffness.

Performance Objectives

In the past, seismic retrofit was primarily applied to achieve public safety, with engineering solutions limited by economic and political considerations. However, with the development of Performance-based earthquake engineering (PBEE), several levels of performance objectives are gradually recognised:

- Public safety only: The goal is to protect human life, ensuring that the structure will not collapse upon its occupants or passersby, and that the structure can be safely exited. Under severe seismic conditions the structure may be a total economic write-off, requiring tear-down and replacement.
- Structure survivability: The goal is that the structure, while remaining safe for exit, may require extensive repair (but not replacement) before it is generally useful or considered safe for occupation. This is typically the lowest level of retrofit applied to bridges.
- Structure functionality: Primary structure undamaged and the structure is undiminished in utility for its primary application. A high level of retrofit, this ensures that any required repairs are only "cosmetic" – for example, minor cracks in plaster, drywall and stucco. This is the minimum acceptable level of retrofit for hospitals.
- Structure unaffected: This level of retrofit is preferred for historic structures of high cultural significance.

Techniques

External Post-tensioning

One of many "earthquake bolts" found throughout period houses in the city of Charleston subsequent to the Charleston earthquake of 1886. They could be tightened and loosened to support the house without having to otherwise demolish the house due to instability. The bolts were directly loosely connected to the supporting frame of the house.

The use of external post-tensioning for new structural systems have been developed in the past decade. Under the PRESS (Precast Seismic Structural Systems), a large-scale U.S./Japan joint research program, unbonded post-tensioning high strength steel tendons have been used to achieve a moment-resisting system that has self-centering capacity. An extension of the same idea for seismic retrofitting has been experimentally tested for seismic retrofit of California bridges under a Caltrans research project and for seismic retrofit of non-ductile reinforced concrete frames. Pre-stressing can increase the capacity of structural elements such as beam, column and beam-column joints. External pre-stressing has been used for structural upgrade for gravity/live loading since the 1970s.

Base Isolators

Base isolation is a collection of structural elements of a building that should substantially decouple the building's structure from the shaking ground thus protecting the building's integrity and enhancing its seismic performance. This earthquake engineering technology, which is a kind of seismic vibration control, can be applied both to a newly designed building and to seismic upgrading of existing structures. Normally, excavations are made around the building and the building is separated from the foundations. Steel or reinforced concrete beams replace the connections to the foundations, while under these, the isolating pads, or base isolators, replace the material removed. While the base isolation tends to restrict transmission of the ground motion to the building, it also keeps the building positioned properly over the foundation. Careful attention to detail is required where the building interfaces with the ground, especially at entrances, stairways and ramps, to ensure sufficient relative motion of those structural elements.

Supplementary Dampers

Supplementary dampers absorb the energy of motion and convert it to heat, thus "damping" resonant effects in structures that are rigidly attached to the ground. In addition to adding energy dissipation capacity to the structure, supplementary damping can reduce the displacement and acceleration demand within the structures. In some cases, the threat of damage does not come from the initial shock itself, but rather from the periodic resonant motion of the structure that repeated ground motion induces. In the practical sense, supplementary dampers act similarly to Shock absorbers used in automotive suspensions.

Tuned Mass Dampers

Tuned mass dampers (TMD) employ movable weights on some sort of springs. These are typically employed to reduce wind sway in very tall, light buildings. Similar designs may be employed to impart earthquake resistance in eight to ten story buildings that are prone to destructive earthquake induced resonances.

Slosh Tank

A slosh tank is a large container of low viscosity fluid (usually water) that may be placed at locations in a structure where lateral swaying motions are significant, such as the roof, and tuned to counter the local resonant dynamic motion. During a seismic (or wind) event the fluid in the tank will slosh back and forth with the fluid motion usually directed and controlled by internal baffles – partitions that prevent the tank itself becoming resonant with the structure. The net dynamic response of the overall structure is reduced due to both the counteracting movement of mass, as well as energy dissipation or vibration damping which occurs when the fluid's kinetic energy is converted to heat by the baffles. Generally the temperature rise in the system will be minimal and is passively cooled by the surrounding air. One Rincon Hill in San Francisco is a skyscraper with a rooftop slosh tank which was designed primarily to reduce the magnitude of lateral swaying motion from wind. A slosh tank is a passive tuned mass damper. In order to be effective the mass of the liquid is usually on the order of 1% to 5% of the mass it is counteracting, and often this requires a significant volume of liquid. In some cases these systems are designed to double as emergency water cisterns for fire suppression.

Active Control System

Very tall buildings ("skyscrapers"), when built using modern lightweight materials, might sway uncomfortably (but not dangerously) in certain wind conditions. A solution to this problem is to include at some upper story a large mass, constrained, but free to move within a limited range, and moving on some sort of bearing system such as an air cushion or hydraulic film. Hydraulic pistons, powered by electric pumps and accumulators, are actively driven to counter the wind forces and natural resonances. These may also, if properly designed, be effective in controlling excessive motion – with or without applied power – in an earthquake. In general, though, modern steel frame high rise buildings are not as subject to dangerous motion as are medium rise (eight to ten story) buildings, as the resonant period of a tall and massive building is longer than the approximately one second shocks applied by an earthquake.

Adhoc Addition of Structural Support/Reinforcement

The most common form of seismic retrofit to lower buildings is adding strength to the existing structure to resist seismic forces. The strengthening may be limited to connections between existing building elements or it may involve adding primary resisting elements such as walls or frames, particularly in the lower stories. Common retrofit measures for unreinforced masonry buildings in the Western United States include the addition of steel frames, the addition of reinforced concrete walls, and in some cases, the addition of base isolation.

Connections between Buildings and their Expansion Additions

Frequently, building additions will not be strongly connected to the existing structure, but simply placed adjacent to it, with only minor continuity in flooring, siding, and roofing. As a result, the addition may have a different resonant period than the original structure, and they may easily detach from one another. The relative motion will then cause the two parts to collide, causing severe structural damage. Seismic modification will either tie the two building components rigidly together so that they behave as a single mass or it will employ dampers to expend the energy from relative motion, with appropriate allowance for this motion, such as increased spacing and sliding bridges between sections.

Exterior Concrete Columns

Historic buildings, made of unreinforced masonry, may have culturally important interior detailing or murals that should not be disturbed. In this case, the solution may be to add a number of steel, reinforced concrete, or poststressed concrete columns to the exterior. Careful attention must be paid to the connections with other members such as footings, top plates, and roof trusses.

Infill Shear Trusses

Shown here is an exterior shear reinforcement of a conventional reinforced concrete dormitory building. In this case, there was sufficient vertical strength in the building columns and sufficient shear strength in the lower stories that only limited shear reinforcement was required to make it earthquake resistant for this location near the Hayward fault.

Massive Exterior Structure

In other circumstances, far greater reinforcement is required. In the structure shown at right — a parking garage over shops — the placement, detailing, and painting of the reinforcement becomes itself an architectural embellishment.

Typical Retrofit Scenario and Solution

Soft-story Failure

This collapse mode is known as *soft story collapse.* In many buildings the ground level is designed for different uses than the upper levels. Low rise residential structures may be built over a parking garage which have large doors on one side. Hotels may have a tall ground floor to allow for a grand entrance or ballrooms. Office buildings may have retail stores on the ground floor with continuous display windows.

Partial failure due to inadequate shear structure at garage level.
Damage in San Francisco due to the Loma Prieta event.

Traditional seismic design assumes that the lower stories of a building are stronger than the upper stories; where this is not the case—if the lower story is less strong than the upper structure—the structure will not respond to earthquakes in the expected fashion. Using modern design methods, it is possible to take a weak lower story into account. Several failures of this type in one large apartment complex caused most of the fatalities in the 1994 Northridge earthquake.

Typically, where this type of problem is found, the weak story is reinforced to make it stronger than the floors above by adding shear walls or moment frames. Moment frames consisting of inverted U bents are useful in preserving lower story garage access, while a lower cost solution may be to use shear walls or trusses in several locations, which partially reduce the usefulness for automobile parking but still allow the space to be used for other storage.

Beam-column Joint Connections

Beam-column joint connections are a common structural weakness in dealing with seismic retrofitting. Prior to the introduction of modern seismic codes in early 1970s, beam-column joints were typically non-engineered or designed. Laboratory testings have confirmed the seismic vulnerability of these poorly detailed and under-designed connections. Failure of beam-column joint connections can typically lead to catastrophic collapse of a frame-building, as often observed in recent earthquakes

For reinforced concrete beam-column joints – various retrofit solutions have been proposed and tested in the past 20 years. Philosophically, the various seismic retrofit strategies can be implemented for reinforced concrete joints. Concrete or steel jacketing have been a popular retrofit technique until the advent of composite materials such as Carbon fiber-reinforced polymer (FRP). Composite materials such as carbon FRP and aramic FRP have been extensively tested for use in seismic retrofit with some success.

One novel technique includes the use of selective weakening of the beam and added external post-tensioning to the joint in order to achieve flexural hinging in the beam, which is more desirable in terms of seismic design.

Corner joint steel reinforcement and high tensile strength rods with grouted anti-burst jacket below.

Widespread weld failures at beam-column joints of low-to-medium rise steel buildings during the Northridge 1994 earthquake for example, have shown the structural defiencies of these 'modern-designed' post-1970s welded moment-resisting connections. A subsequent SAC research project has documented, tested and proposed several retrofit solutions for these welded steel moment-resisting connections. Various retrofit solutions have been developed for these welded joints – such as a) weld strengthening and b) addition of steel haunch or 'dog-bone' shape flange.

Following the Northridge earthquake, a number of steel moment -frame buildings were found to have experienced brittle fractures of beam to column connections. Discovery of these unanticipated brittle fractures of framing connections was alarming to engineers and the building industry. Starting in the 1960s, engineers began to regard welded steel moment-frame buildings as being among the most ductile systems contained in the building code. Many engineers believed that steel moment-frame buildings were essentially invulnerable to earthquake induced damage and thought that should damage occur, it would be limited to ductile yielding of members and connections. Observation of damage sustained by buildings in the 1994 Northridge earthquake indicated that contrary to the intended behavior, in many cases, brittle fractures initiated within the connections at very low levels of plastic demand. In September, 1994, The SAC joint Venture, AISC, AISI, and NIST jointly convened an international workshop in Los Angeles to coordinate the efforts of various participants and to lay the foundation for systematic investigation and resolution of the problem. In September 1995 the SAC Joint Venture entered into a contractual agreement with FEMA to conduct Phase II of the SAC Steel project. Under Phase II, SAC continued its extensive problem-focused study of the performance of moment resisting steel frames and connections of various configurations, with the ultimate goal of developing seismic design criteria for steel construction. As a result of these studies it is now known that the typical moment-resisting connection detail employed in steel moment frame construction prior to the 1994 Northridge earthquake had a number of features that rendered it inherently susceptible to brittle fracture.

Shear Failure within Floor Diaphragm

Floors in wooden buildings are usually constructed upon relatively deep spans of wood, called joists, covered with a diagonal wood planking or plywood to form a subfloor upon which the finish floor surface is laid. In many structures these are all aligned in the same direction. To prevent the beams from tipping over onto their side, blocking is used at each end, and for additional stiffness, blocking or diagonal wood or metal bracing may be placed between beams at one or more points in their spans. At the outer edge it is typical to use a single depth of blocking and a perimeter beam overall.

If the blocking or nailing is inadequate, each beam can be laid flat by the shear forces applied to the building. In this position they lack most of their original strength and the structure may further collapse. As part of a retrofit the blocking may be doubled, especially at the outer edges of the building. It may be appropriate to add additional nails between the sill plate of the perimeter wall erected upon the floor diaphragm, although this will require exposing the sill plate by removing interior plaster or exterior siding. As the sill plate may be quite old and dry and substantial nails must be used, it may be necessary to pre-drill a hole for the nail in the old wood to avoid splitting. When the wall is opened for this purpose it may also be appropriate to tie vertical wall elements into the foundation using specialty connectors and bolts glued with epoxy cement into holes drilled in the foundation.

Sliding off Foundation and "Cripple Wall" Failure

House slid off of foundation.

Low cripple wall collapse and detachment of structure from concrete stairway.

Single or two story wood-frame domestic structures built on a perimeter or slab foundation are relatively safe in an earthquake, but in many structures built before 1950 the sill plate that sits between the concrete foundation and the floor diaphragm (perimeter foundation) or studwall (slab foundation) may not be sufficiently bolted in. Additionally, older attachments (without substantial corrosion-proofing) may have corroded to a point of weakness. A sideways shock can slide the building entirely off of the foundations or slab.

Often such buildings, especially if constructed on a moderate slope, are erected on a platform connected to a perimeter foundation through low stud-walls called "cripple wall" or *pin-up*. This low wall structure itself may fail in shear or in its connections to itself at the corners, leading to the building moving diagonally and collapsing the low walls. The likelihood of failure of the pin-up can be reduced by ensuring that the corners are well reinforced in shear and that the shear panels are well connected to each other through the corner posts. This requires structural grade sheet plywood, often treated for rot resistance. This grade of plywood is made without interior unfilled knots and with more, thinner layers than common plywood. New buildings designed to resist earthquakes will typically use OSB (oriented strand board), sometimes with metal joins between panels, and with well attached stucco covering to enhance its performance. In many modern tract homes, especially those built upon expansive (clay) soil the building is constructed upon a single and relatively thick monolithic slab, kept in one piece by high tensile rods that are stressed after the slab has set. This poststressing places the concrete under compression – a condition under which it is extremely strong in bending and so will not crack under adverse soil conditions.

Multiple Piers in Shallow Pits

Some older low-cost structures are elevated on tapered concrete pylons set into shallow pits, a method frequently used to attach outdoor decks to existing buildings. This is seen in conditions of damp soil, especially in tropical conditions, as it leaves a dry ventilated space under the house, and in far northern conditions of permafrost (frozen mud) as it keeps the building's warmth from destabilizing the ground beneath. During an earthquake, the pylons may tip, spilling the building to the ground. This can be overcome by using deep-bored holes to contain cast-in-place reinforced pylons, which are then secured to the floor panel at the corners of the building. Another technique is to add sufficient diagonal bracing or sections of concrete shear wall between pylons.

Reinforced Concrete Column Burst

Reinforced concrete columns typically contain large diameter vertical rebar (reinforcing bars) arranged in a ring, surrounded by lighter-gauge hoops of rebar. Upon analysis of failures due to earthquakes, it has been realized that the weakness was not in the vertical bars, but rather in inadequate strength and quantity of hoops. Once the integrity of the hoops is breached, the vertical rebar can flex outward, stressing the central column of concrete. The concrete then simply crumbles into small pieces, now unconstrained by the surrounding rebar. In new construction a greater amount of hoop-like structures are used.

Jacketed and grouted column on left, unmodified on right.

One simple retrofit is to surround the column with a jacket of steel plates formed and welded into a single cylinder. The space between the jacket and the column is then filled with concrete, a process called grouting. Where soil or structure conditions require such additional modification, additional pilings may be driven near the column base and concrete pads linking the pilings to the pylon are fabricated at or below ground level. In the example shown not all columns needed to be modified to gain sufficient seismic resistance for the conditions expected.

Reinforced Concrete Wall Burst

Concrete walls are often used at the transition between elevated road fill and overpass structures. The wall is used both to retain the soil and so enable the use of a shorter span and also to transfer the weight of the span directly downward to footings in undisturbed soil. If these walls are inadequate they may crumble under the stress of an earthquake's induced ground motion.

One form of retrofit is to drill numerous holes into the surface of the wall, and secure short L-shaped sections of rebar to the surface of each hole with epoxy adhesive. Additional vertical and horizontal rebar is then secured to the new elements, a form is erected, and an additional layer of concrete is poured. This modification may be combined with additional footings in excavated trenches and additional support ledgers and tie-backs to retain the span on the bounding walls.

Brick Wall Resin and Glass Fiber Reinforcement

Brick building structures have been reinforced with coatings of glass fiber and appropriate resin (epoxy or polyester). In lower floors these may be applied over entire exposed surfaces, while in upper floors this may be confined to narrow areas around window and door openings. This application provides tensile strength that stiffens the wall against bending away from the side with the application. The efficient protection of an entire building requires extensive analysis and engineering to determine the appropriate locations to be treated.

Lift

Where moist or poorly consolidated alluvial soil interfaces in a "beach like" structure against underlying firm material, seismic waves traveling through the alluvium can be amplified, just as are water waves against a sloping beach. In these special conditions, vertical accelerations up to twice the force of gravity have been measured. If a building is not secured to a well-embedded foundation it is possible for the building to be thrust from (or with) its foundations into the air, usually with severe damage upon landing. Even if it is well-founded, higher portions such as upper stories or roof structures or attached structures such as canopies and porches may become detached from the primary structure.

Good practices in modern, earthquake-resistant structures dictate that there be good vertical connections throughout every component of the building, from undisturbed or engineered earth to foundation to sill plate to vertical studs to plate cap through each floor and continuing to the roof structure. Above the foundation and sill plate the connections are typically made using steel strap or sheet stampings, nailed to wood members using special hardened high-shear strength nails, and heavy angle stampings secured with through bolts, using large washers to prevent pull-through. Where inadequate bolts are provided between the sill plates and a foundation in existing construction (or are not trusted due to possible corrosion), special clamp plates may be added, each of which is secured to the foundation using expansion bolts inserted into holes drilled in an exposed face of concrete. Other members must then be secured to the sill plates with additional fittings.

Soil

One of the most difficult retrofits is that required to prevent damage due to soil failure. Soil failure can occur on a slope, a slope failure or landslide, or in a flat area due to liquefaction of water-saturated sand and/or mud. Generally, deep pilings must be driven into stable soil (typically hard mud or sand) or to underlying bedrock or the slope must be stabilized. For buildings built atop previous landslides the practicality of retrofit may be limited by economic factors, as it is not practical to stabilize a large, deep landslide. The likelihood of landslide or soil failure may also depend upon seasonal factors, as the soil may be more stable at the beginning of a wet season than at the beginning of the dry season. Such a "two season" *Mediterranean climate* is seen throughout California.

In some cases, the best that can be done is to reduce the entrance of water runoff from higher, stable elevations by capturing and bypassing through channels or pipes, and to drain water infiltrated directly and from subsurface springs by inserting horizontal perforated tubes. There are numerous locations in California where extensive developments have been built atop archaic landslides, which have not moved in historic times but which (if both water-saturated and shaken by an earthquake) have a high probability of moving *en masse*, carrying entire sections of suburban development to new locations. While the most modern of house structures (well tied to monolithic concrete foundation slabs reinforced with post tensioning cables) may survive such movement largely intact, the building will no longer be in its proper location.

Utility Pipes and Cables: Risks

Natural gas and propane supply pipes to structures often prove especially dangerous during and after earthquakes. Should a building move from its foundation or fall due to cripple wall collapse, the ductile iron pipes transporting the gas within the structure may be broken, typically at the location of threaded joints. The gas may then still be provided to the pressure regulator from higher pressure lines and so continue to flow in substantial quantities; it may then be ignited by a nearby source such as a lit pilot light or arcing electrical connection.

There are two primary methods of automatically restraining the flow of gas after an earthquake, installed on the low pressure side of the regulator, and usually downstream of the gas meter.

- A caged metal ball may be arranged at the edge of an orifice. Upon seismic shock, the ball will roll into the orifice, sealing it to prevent gas flow. The ball may later be reset by the use of an external magnet. This device will respond only to ground motion.
- A flow-sensitive device may be used to close a valve if the flow of gas exceeds a set threshold (very much like an electrical circuit breaker). This device will operate independently of seismic motion, but will not respond to minor leaks which may be caused by an earthquake.

It appears that the most secure configuration would be to use one of each of these devices in series.

Tunnels

Unless the tunnel penetrates a fault likely to slip, the greatest danger to tunnels is a landslide blocking an entrance. Additional protection around the entrance may be applied to divert any falling material (similar as is done to divert snow avalanches) or the slope above the tunnel may be stabilized in some way. Where only small- to medium-sized rocks and boulders are expected to fall, the entire slope may be covered with wire mesh, pinned down to the slope with metal rods. This is also a common modification to highway cuts where appropriate conditions exist.

Underwater Tubes

The safety of underwater tubes is highly dependent upon the soil conditions through which the tunnel was constructed, the materials and reinforcements used, and the maximum predicted earthquake expected, and other factors, some of which may remain unknown under current knowledge.

BART Tube

A tube of particular structural, seismic, economic, and political interest is the BART (Bay Area Rapid Transit) transbay tube. This tube was constructed at the bottom of San Francisco Bay through an innovative process. Rather than pushing a shield through the soft bay mud, the tube was constructed on land in sections. Each section consisted of two inner train tunnels of circular cross section, a central access tunnel of rectangular cross section, and an outer oval shell encompassing the three inner tubes. The intervening space was filled with concrete. At the bottom of the bay a trench was excavated and a flat bed of crushed stone prepared to receive the tube sections. The sections were then floated into place and sunk, then joined with bolted connections to previously-placed sections. An overfill was then placed atop the tube to hold it down. Once completed from San Francisco to Oakland, the tracks and electrical components were installed. The predicted response of the tube during a major earthquake was likened to be as that of a string of (cooked) spaghetti in a bowl of gelatin dessert. To avoid overstressing the tube due to differential movements at each end, a sliding slip joint was included at the San Francisco terminus under the landmark Ferry Building.

The engineers of the construction consortium PBTB (Parsons Brinckerhoff-Tudor-Bechtel) used the best estimates of ground motion available at the time, now known to be insufficient given modern computational analysis methods and geotechnical knowledge. Unexpected settlement of the tube has reduced the amount of slip that can be accommodated without failure. These factors have resulted in the slip joint being designed too short to ensure survival of the tube under possible (perhaps even likely) large earthquakes in the region. To correct this deficiency the slip joint must be extended to allow for additional movement, a modification expected to be both expensive and technically and logistically difficult. Other retrofits to the BART tube include vibratory consolidation of the tube's overfill to avoid potential liquefying of the overfill, which has now been completed. (Should the overfill fail there is a danger of portions of the tube rising from the bottom, an event which could potentially cause failure of the section connections.)

Failure Modes of Bridge Retrofit

Expansion Rockers

Many short bridge spans are statically anchored at one end and attached to rockers at the other. This rocker gives vertical and transverse support while allowing the bridge span to expand and contract with temperature changes. The change in the length of the span is accommodated over a gap in the roadway by comb-like expansion joints. During severe ground motion, the rockers may jump from their tracks or be moved beyond their design limits, causing the bridge to unship from its resting point and then either become misaligned or fail completely. Motion can be constrained by adding ductile or high-strength steel restraints that are friction-clamped to beams and designed to slide under extreme stress while still limiting the motion relative to the anchorage.

Deck Rigidity

Additional diagonals were inserted under both decks of this bridge.

Suspension bridges may respond to earthquakes with a side-to-side motion exceeding that which was designed for wind gust response. Such motion can cause fragmentation of the road surface, damage to bearings, and plastic deformation or breakage of components. Devices such as hydraulic dampers or clamped sliding connections and additional diagonal reinforcement may be added.

Lattice Girders, Beams and Ties

Obsolete riveted lattice members

Lattice girders consist of two "I"-beams connected with a criss-cross lattice of flat strap or angle stock. These can be greatly strengthened by replacing the open lattice with plate members. This is usually done in concert with the replacement of hot rivets with bolts.

Bolted plate lattice replacement, forming box members.

Hot Rivets

Many older structures were fabricated by inserting red-hot rivets into pre-drilled holes; the soft rivets are then peened using an air hammer on one side and a bucking bar on the head end. As these cool slowly, they are left in an annealed (soft) condition, while the plate, having been hot rolled and quenched during manufacture, remains relatively hard. Under extreme stress the hard plates can shear the soft rivets, resulting in failure of the joint.

The solution is to burn out each rivet with an oxygen torch. The hole is then prepared to a precise diameter with a reamer. A special *locator bolt*, consisting of a head, a shaft matching the reamed hole, and a threaded end is inserted and retained with a nut, then tightened with a wrench. As the bolt has been formed from an appropriate high-strength alloy and has also been heat-treated, it is not subject to either the plastic shear failure typical of hot rivets nor the brittle fracture of ordinary bolts. Any partial failure will be in the plastic flow of the metal secured by the bolt; with proper engineering any such failure should be non-catastrophic.

Fill and Overpass

Elevated roadways are typically built on sections of elevated earth fill connected with bridge-like segments, often supported with vertical columns. If the soil fails where a bridge terminates, the bridge may become disconnected from the rest of the roadway and break away. The retrofit for this is to add additional reinforcement to any supporting wall, or to add deep caissons adjacent to the edge at each end and connect them with a supporting beam under the bridge.

Another failure occurs when the fill at each end moves (through resonant effects) in bulk, in opposite directions. If there is an insufficient founding shelf for the overpass, then it may fall. Additional shelf and ductile stays may be added to attach the overpass to the footings at one or both ends. The stays, rather than being fixed to the beams, may instead be clamped to them. Under moderate loading, these keep the overpass centered in the gap so that it is less likely to slide off its founding shelf at one end. The ability for the fixed ends to slide, rather than break, will prevent the complete drop of the structure if it should fail to remain on the footings.

Viaducts

Large sections of roadway may consist entirely of viaduct, sections with no connection to the earth other than through vertical columns. When concrete columns are used, the detailing is critical. Typical failure may be in the toppling of a row of columns due either to soil connection failure or to insufficient cylindrical wrapping with rebar. Both failures were seen in the 1995 Great Hanshin earthquake in Kobe, Japan, where an entire viaduct, centrally supported by a single row of large columns, was laid down to one side. Such columns are reinforced by excavating to the foundation pad, driving additional pilings, and adding a new, larger pad, well connected with rebar alongside or into the column. A column with insufficient wrapping bar, which is prone to burst and then hinge at the bursting point, may be completely encased in a circular or elliptical jacket of welded steel sheet and grouted.

Cypress Freeway viaduct collapse. Note failure of inadequate anti-burst wrapping and lack of connection between upper and lower vertical elements.

Sometimes viaducts may fail in the connections between components. This was seen in the failure of the Cypress Freeway in Oakland, California, during the Loma Prieta earthquake. This viaduct was a two-level structure, and the upper portions of the columns were not well connected to the lower portions that supported the lower level; this caused the upper deck to collapse upon the lower deck. Weak connections such as these require additional external jacketing – either through external steel components or by a complete jacket of reinforced concrete, often using stub connections that are glued (using epoxy adhesive) into numerous drilled holes. These stubs are then connected to additional wrappings, external forms (which may be temporary or permanent) are erected, and additional concrete is poured into the space. Large connected structures similar to the Cypress Viaduct must also be properly analyzed in their entirety using dynamic computer simulations.

Residential Retrofit

Side-to-side forces cause most earthquake damage. Bolting of the mudsill to the foundation and application of plywood to cripple walls are a few basic retrofit techniques which homeowners may apply to wood-framed residential structures to mitigate the effects of seismic activity. The City of San Leandro created guidelines for these procedures, as outlined in the following pamphlet. Public awareness and initiative are critical to the retrofit and preservation of existing building stock, and such efforts as those of the Association of Bay Area Governments are instrumental in providing informational resources to seismically active communities.

Wood Frame Structure

Most houses in North America are wood-framed structures. Wood is one of the best materials for earthquake-resistant construction since it is lightweight and more flexible than masonry. It is easy to work with and less expensive than steel, masonry, or concrete. In older homes the most significant weaknesses are the connection from the wood-framed walls to the foundation and the relatively weak "cripple-walls." (Cripple walls are the short wood walls that extend from the top of the foundation to the lowest floor level in houses that have raised floors.) Adding connections from the base of the wood-framed structure to the foundation is almost always an important part of a seismic retrofit. Bracing the cripple-walls to resist side-to-side forces is essential in houses with cripple walls; bracing is usually done with plywood. Oriented strand board (OSB) does not perform as consistently as plywood, and is not the favored choice of retrofit designers or installers.

Retrofit methods in older wood-frame structures may consist of the following:

- The lowest plate rails of walls (usually called "mudsills" or "foundation sills" in North America) are bolted to a continuous foundation, or secured with rigid metal connectors bolted to the foundation so as to resist side-to-side forces.
- *Cripple walls* are braced with plywood.
- Selected vertical elements (typically the posts at the ends of plywood wall bracing panels) are connected to the foundation. These connections are intended to prevent the braced walls from rocking up and down when subjected to back-and-forth forces at the top of the braced walls, not to resist the wall or house "jumping" off the foundation (which almost never occurs).
- In two story buildings using "platform framing" (sometimes called "western" style construction, where walls are progressively erected upon the lower

story's upper diaphragm, unlike "eastern" or *balloon framing*), the upper walls are connected to the lower walls with tension elements. In some cases, connections may be extended vertically to include retention of certain roof elements. This sort of strengthening is usually very costly with respect to the strength gained.

- Vertical posts are secured to the beams or other members they support. This is particularly important where loss of support would lead to collapse of a segment of a building. Connections from posts to beams cannot resist appreciable side-to-side forces; it is much more important to strengthen around the perimeter of a building (bracing the cripple-walls and supplementing foundation-to-wood-framing connections) than it is to reinforce post-to-beam connections.

Wooden framing is efficient when combined with masonry, if the structure is properly designed. In Turkey, the traditional houses (bagdadi) are made with this technology. In El Salvador, wood and bamboo are used for residential construction.

Reinforced and Unreinforced Masonry

In many parts of developing countries such as Pakistan, Iran and China, unreinforced or in some cases reinforced masonry is the predominantly form of structures for rural residential and dwelling. Masonry was also a common construction form in the early part of the 20th century, which implies that a substantial number of these at-risk masonry structures would have significant heritage value. Masonry walls that are not reinforced are especially hazardous. Such structures may be more appropriate for replacement than retrofit, but if the walls are the principal load bearing elements in structures of modest size they may be appropriately reinforced. It is especially important that floor and ceiling beams be securely attached to the walls. Additional vertical supports in the form of steel or reinforced concrete may be added.

In the western United States, much of what is seen as masonry is actually brick or stone veneer. Current construction rules dictate the amount of *tie–back* required, which consist of metal straps secured to vertical structural elements. These straps extend into mortar courses, securing the veneer to the primary structure. Older structures may not secure this sufficiently for seismic safety. A weakly secured veneer in a house interior (sometimes used to face a fireplace from floor to ceiling) can be especially dangerous to occupants. Older masonry chimneys are also dangerous if they have substantial vertical extension above the roof. These are prone to breakage at the roofline and may fall into the house in a single large piece. For retrofit, additional supports may be added; however, it is extremely expensive to strengthen an existing masonry chimney to conform with contemporary design standards. It is best to simply remove the extension and replace it with lighter materials, with special metal flue replacing the flue tile and a wood structure replacing the masonry. This may be matched against existing brickwork by using very thin veneer (similar to a tile, but with the appearance of a brick).

SEISMIC METAMATERIAL

A seismic metamaterial, is a metamaterial that is designed to counteract the adverse effects of seismic waves on artificial structures, which exist on or near the surface of the earth. Current designs of seismic metamaterials utilize configurations of boreholes, trees or proposed underground resonators to act as a large scale material. Experiments have observed both reflections and bandgap attenuation from artificially induced seismic waves. These are the first experiments to verify that seismic metamaterials can be measured for frequencies below 100 Hz, where damage from Rayleigh waves is the most harmful to artificial structures.

The Mechanics of Seismic Waves

More than a million earthquakes are recorded each year, by a worldwide system of earthquake detection stations. The propagation velocity of the seismic waves depends on density and elasticity of the earth materials. In other words, the speeds of the seismic waves vary as they travel through different materials in the earth. The two main components of a seismic event are body waves and surface waves. Both of these have different modes of wave propagation.

Seismic Cloaking

Computations showed that seismic waves traveling toward a building, could be directed around the building, leaving the building unscathed, by using *seismic metamaterials*. The very long wavelengths of earthquake waves would be shortened as they interact with the metamaterials; the waves would pass around the building so as to arrive in phase as the earthquake wave proceeded, as if the building was not there. The mathematical models produce the regular pattern provided by Metamaterial cloaking. This method was first understood with electromagnetic cloaking metamaterials - the electromagnetic energy is in effect directed around an object, or hole, and protecting buildings from seismic waves employs this same principle.

Giant polymer-made split ring resonators combined with other metamaterials are designed to couple at the seismic wavelength. Concentric layers of this material would be stacked, each layer separated by an elastic medium. The design that worked is ten layers of six different materials, which can be easily deployed in building foundations. As of 2009, the project is still in the design stage.

Electromagnetics Cloaking Principles for Seismic Metamaterials

For seismic metamaterials to protect surface structures, the proposal includes a layered structure of metamaterials, separated by elastic plates in a cylindrical configuration. A prior simulation showed that it is possible to create concealment from electromagnetic radiation with concentric, alternating layers of electromagnetic metamaterials. That study was is in contrast to concealment by inclusions in a split ring resonator designed as an anisotropic metamaterial.

The configuration can be viewed as alternating layers of "homogeneous isotropic dielectric material" A. with "homogeneous isotropic dielectric material" B. Each dielectric material is much thinner than the radiated wavelength. As a whole, such structure is an anisotropic medium. The layered dielectric materials surround an "infinite conducting cylinder". The layered dielectric materials radiate outward, in a concentric fashion, and the cylinder is encased in the first layer. The other layers alternate and surround the previous layer all the way to the first layer. Electromagnetic wave scattering was calculated and simulated for the layered (metamaterial) structure and the split-ring resonator anistropic metamaterial, to show the effectiveness of the layered metamaterial.

Acoustic Cloaking Principles for Seismic Metamaterials

The theory and ultimate development for the *seismic metamaterial* is based on coordinate transformations achieved when concealing a small cylindrical object with electromagnetic waves. This was followed by an analysis of acoustic cloaking, and whether or not coordinate transformations could be applied to artificially fabricated acoustic materials.

Applying the concepts used to understand electromagnetic materials to material properties in other systems shows them to be closely analogous. Wave vector, wave impedance, and direction of power flow are universal. By understanding how permittivity and permeability control these components of wave propagation, applicable analogies can be used for other material interactions.

In most instances, applying coordinate transformation to engineered artificial elastic media is not possible. However, there is at least one special case where there is a direct equivalence between electromagnetics and elastodynamics. Furthermore, this case appears practically useful. In two dimensions, isotropic acoustic media and isotropic electromagnetic media are exactly equivalent. Under these conditions, the isotropic characteristic works in anisotropic media as well.

It has been demonstrated mathematically that the 2D Maxwell equations with normal incidence apply to 2D acoustic equations when replacing the electromagnetic parameters with the following acoustic parameters: pressure, vector fluid velocity, fluid mass density and the fluid bulk modulus. The compressional wave solutions used in the electromagnetic cloaking are transferred to material fluidic solutions where fluid motion is parallel to the wavevector. The computations then show that coordinate transformations can be applied to acoustic media when restricted to normal incidence in two dimensions.

Next the electromagnetic cloaking shell is referenced as an exact equivalence for a simulated demonstration of the acoustic cloaking shell. Bulk modulus and mass density determine the spatial dimensions of the cloak, which can bend any incident wave around the center of the shell. In a simulation with perfect conditions, because it is easier to demonstrate the principles involved, there is zero scattering in any direction.

The Seismic Cloak

However, it can be demonstrated through computation and visual simulation that the waves are in fact dispersed around the location of the building. The frequency range of this capability is shown to have no limitation regarding the radiated frequency. The cloak itself demonstrates no forward or back scattering, hence, the seismic cloak becomes an effective medium.

ENDURANCE TIME METHOD

Endurance time (ET) method is a dynamic structural analysis procedure for seismic assessment of structures. In this procedure, an intensifying dynamic excitation is used as the loading function. Endurance time method is a time-history based dynamic analysis procedure. An estimate of the structural response at different equivalent seismic intensity levels is obtained in a single response history analysis. This method has applications in seismic assessment of various structural types and in different areas of earthquake engineering.

Endurance time (ET) method is a dynamic structural analysis procedure in which intensifying dynamic excitation is used as the loading function. An estimate of structural response and/or performance at the entire seismic intensity range of interest is obtained in each response history analysis. The concept of endurance time analysis is similar to the exercise test applied in medicine. Similar concept has also been extended to applications in the analysis of offshore platforms under water waves.

ET Excitation Functions

ET excitation functions are generated by using numerical optimization methods. ET excitation functions are publicly available through internet websites. ET excitation functions are categorized into four generations as follows:

- First generation of ET excitation functions (ETEFs) are essentially a filtered and profiled white noise. These were used for demonstrating the concept of ET and have limited practical significance.
- Second-generation ETEFs incorporate response spectrum matching. These ETEFs produce numerically significant analysis results.
- Third-generation ETEFs are optimized in nonlinear range. These ETEFs deliver improved analysis performance.

- Fourth-generation ETEFs are optimized to include duration consistency.

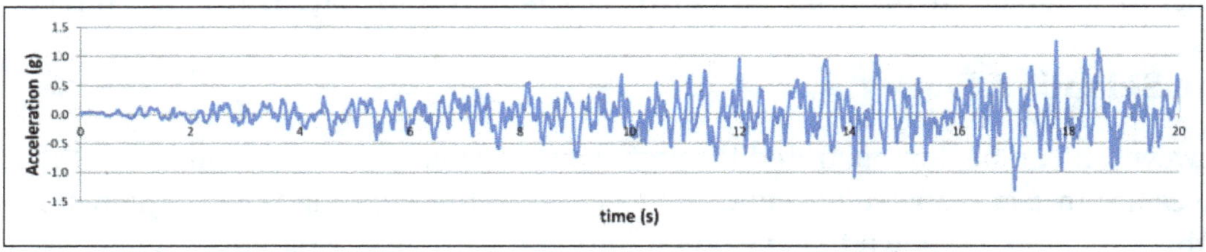

A typical endurance time excitation function (ETA20f03 record).

Application Areas in Earthquake Engineering

Endurance time method has been applied in the following areas of earthquake engineering:

- Nonlinear dynamic analysis of structures.
- Seismic evaluation of jacket-type offshore platforms.
- Optimal damper placement in framed buildings.
- Seismic assessment of structures.
- Performance-based design.
- Structural optimization.
- Multi-component seismic analysis.
- Soil–structure interaction.
- Soil-pile-superstructure interaction.
- Liquid–structure interaction.
- Dam engineering.
- Bridge engineering.
- Seismic rehabilitation.
- Collapse analysis.

Structural Type Applications

ET method has been applied in seismic assessment of the following structural types:

- Single degree of freedom systems.
- Moment and braced steel frames.

- Concrete frames.
- Bridges.
- Gravity dams.
- Arch dams.
- Shell structures.
- Steel tanks.
- Offshore structures.

Advantages of ET Method

Major advantages of the endurance time method are as follows:

- ET significantly reduces the computational demand required for performing a standard response history analysis of structures for seismic assessment, especially when response at multiple levels of intensity is to be considered.
- ET is applicable in a wide range of seismic assessment problems and provides a generic approach for the seismic analysis of a wide range of structural types.
- ET method is reasonably simple and sensible when a realistic dynamic analysis of a complex structure is required.

Limitations of ET Method

Major limitations of the endurance time method are as follows:

- ET is an approximate method for predicting the structural response.
- The production of usable ETEFs that are applicable in a particular situation can be complicated.
- The procedure is still under development and sufficient background information may not be available for specific applications.

VISCOELASTIC DAMPER

Viscoelastic damper is a passive energy dissipation device. This damper converts a portion of mechanical energy into heat energy. The medium in which the transfer of energy takes place is viscoelastic material. Viscoelastic dampers are known to get activated at lower displacements, hence optimum against governing wind forces too. Visco elastic dampers are known for providing restoring forces. As visco elastic damper follows a linear behavior, it makes the modeling simpler. Viscoelastic damper does not require any external source for providing energy.

There are basically three ways of employing viscoelastic material as a damper medium.

- Single Layer Viscoelastic damper: Here the damping is achieved by extensional deformation of viscoelastic layer.
- Double Layer Viscoelastic damper: Here the viscoelastic material is placed between two constraints layers. Damping achieved is mostly due to shear deformation rather than extensional deformation.

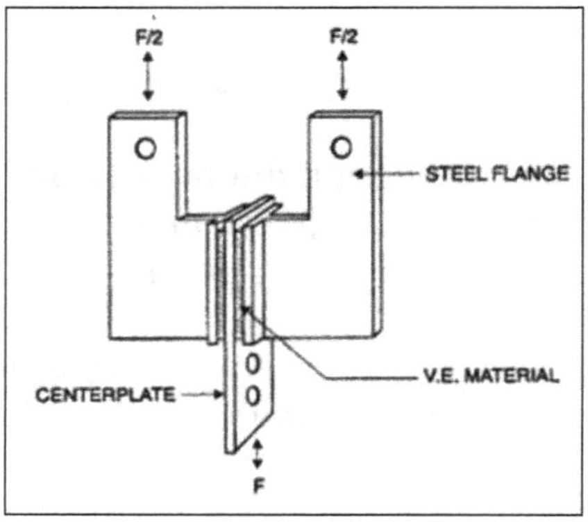

Sketch of Viscoelastic damper.

- Sandwiched viscoelastic damper: Here all the observed deformation is due to shear. This type of damper is more efficient when large amount of energies are to be damped.

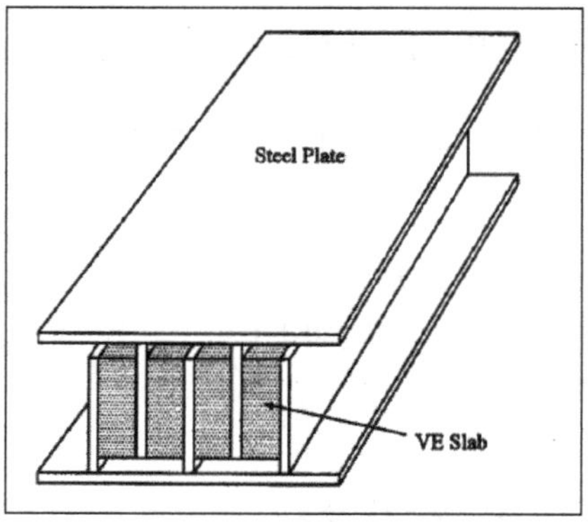

Sketch of Sandwiched viscoelastic damper.

Viscoelastic Materials

A viscoelastic material is characterized by possessing both viscous and elastic behavior. A purely elastic material is one in which all the energy stored in the sample during loading is returned when the load is removed. As a result, the stress and strain curves for elastic materials move completely in phase. For elastic materials, Hooke's Law applies where the stress is proportional to the strain, and the modulus is defined at the ratio of stress to strain. A complete opposite to an elastic material is a purely viscous material, also shown in figure. This type of material does not return any of the energy stored during loading. All the energy is lost as "pure damping" once the load is removed. In this case, the stress is proportional to the rate of the strain, and the ratio of stress to strain rate is known as viscosity, μ. These materials have no stiffness component, only damping. While in case of viscoelastic materials, Some of the energy stored in a viscoelastic system is recovered upon removal of the load and the remainder is dissipated in the form of heat. Most important of these include temperature and frequency effects. These materials exist in various unique states or "phases" over the broad temperature and frequency ranges in which they are used. These regions are typically referred to as the Glassy, Transition, Rubbery, and Flow Regions.

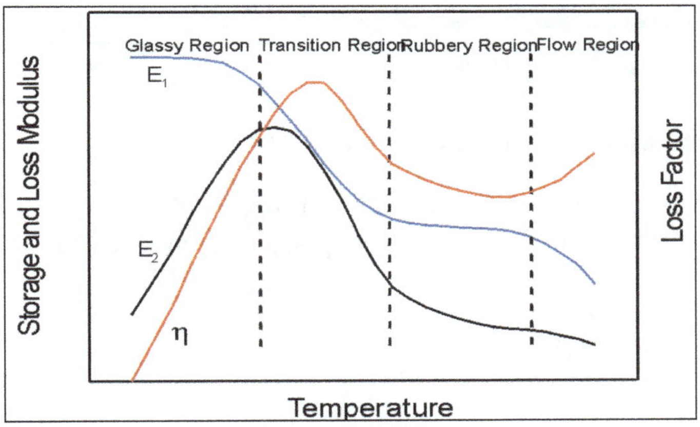

Phases of Viscoelastic material.

FLUID VISCOUS DAMPER

Fluid Viscous Dampers (FVD) are passive control systems of structural response. These systems have the ability to transmit developed forces according to the request of the structural response. Passive control devices dissipate energy in the structure, but cannot increase it. Because of their great ability to return a building to its original position after an earthquake, they are increasingly used in the bracing structures in civil engineering in general and in the metallic high-rise structures in particular. The additional cost of the damper is typically offset by the savings in the steel weight and foundation concrete volume.

Fluid viscous dampers were initially used in the military and aerospace industry. FVD typically consist of a piston head with orifices contained in a cylinder filled with a highly viscous fluid, usually a compound of silicone or a similar type of oil. Energy is dissipated in the damper by fluid orifice when the piston head moves through the fluid. The fluid in the cylinder is nearly incompressible, and when the damper is subjected to a compressive force, the fluid volume inside the cylinder is decreased as a result of the piston rod area movement. A decrease in volume results in a restoring force. This undesirable force is prevented by using an accumulator. An accumulator works by collecting the volume of fluid that is displaced by the piston rod and storing it in the makeup area. As the rod retreats, a vacuum that has been created will draw the fluid out.

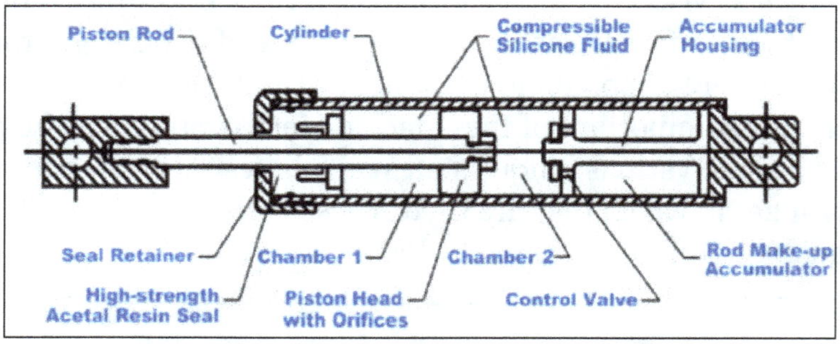

Fluid viscous dampers (FVD).

Characteristics of Fluid Viscous Dampers

FVD are characterised by a resistance force F. It depends on the velocity of movement, the fluid viscosity and the orifices size of the piston. The value of P is given by the relationship:

$$P = C_d \cdot \left(\dot{u}_d\right)^{\alpha} \cdot \sin\left(\dot{u}_d\right)$$

With

$$u_d(t) = u_0 \cdot \sin(\omega \cdot t)$$

where,

- $\dot{u}_d$ is the velocity between two ends of the damper;
- C_d is the damping constant;
- u_0 is the amplitude of the displacement, ω is the loading frequency, and t is time;
- α is an exponent which depends on the viscosity properties of the fluid and the piston.

The value of the constant α may be less than or equal to 1. Figure, show the force velocity and the force displacement relationships for three different types of FVD. They characterise the behaviour of the viscous damper. With $\alpha = 1$ the device is called linear viscous damper and for $\alpha < 1$ non-linear FVD which is effective in minimising high velocity shocks. Damper with $\alpha > 1$ has not been seen often in practical application. The non-linear damper can give a larger damping force than the two other types.

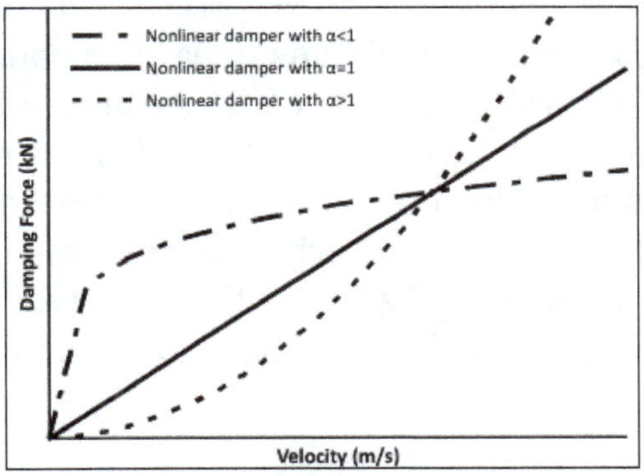

Force velocity relationship of FVD.

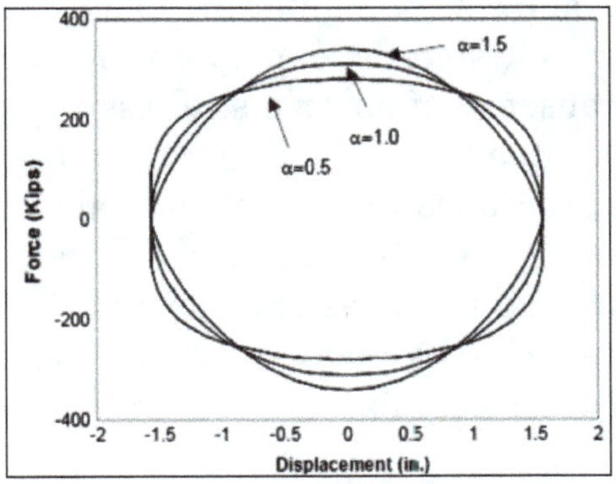

Force displacement relationship of FVD.

Figure above shows that the plot has different shapes for the different values of α. At the frequency of loading used to create the loops enclosed areas for the different damper are all equal, but the values of the damping coefficient are all different.

The resisting force in the FVD, P, can be described by the following equation:

$$P = K_1 \cdot u_d + C_d \cdot \frac{du_d}{dt}$$

where K_1 is the storage stiffness and C is the damping coefficient given by

$$C_d = \frac{K_2}{\omega}$$

where K_2 is the loss stiffness. In $P = K_1 \cdot u_d + C_d \cdot \frac{du_d}{dt}$ the first term represents the force due to the stiffness of the damper, which is in phase with the motion, and the second term represents the force due to the viscosity of the damper, which is 90° out of phase with the motion. Figure below plots the force-displacement relationship for the first and second terms of $P = K_1 \cdot u_d + C_d \cdot \frac{du_d}{dt}$, while (c) plots the total force. Figure below shows the structure's behaviour without dampers.

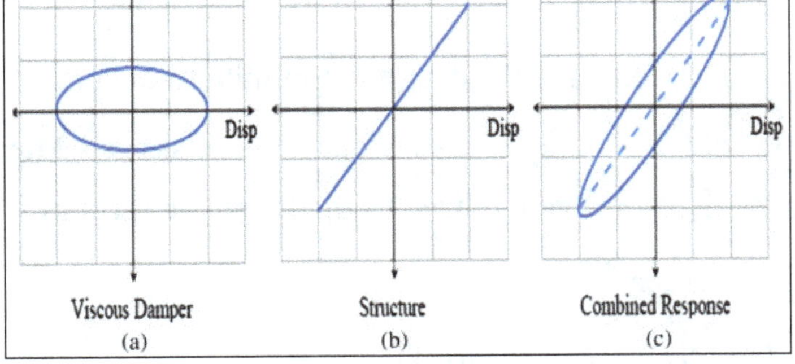

Hysteretic curve of FVD.

FVD allow very significant energy dissipation where the stress-strains diagram shows a hysteretic loop approaching an ellipse for a pure viscous linear behaviour. The absence of storage stiffness makes the natural frequency of the structure incorporated with the damper remains the same. This advantage will simplify the design procedure with supplemental viscous devices. However if the damper develops restoring force the loop will be changed in figure above. It turns from viscous behaviour to viscoelastic behaviour. The maximum energy amount that this type of damper can dissipate in a very short time is only limited by the thermal capacity of lead and steel tube.

Analytical Model of the Fluid Viscous Damper

Fluid viscous dampers exhibit a viscoelastic behaviour, which can be best predicted with the Kelvin and Maxwell models for linear and non linear models respectively.

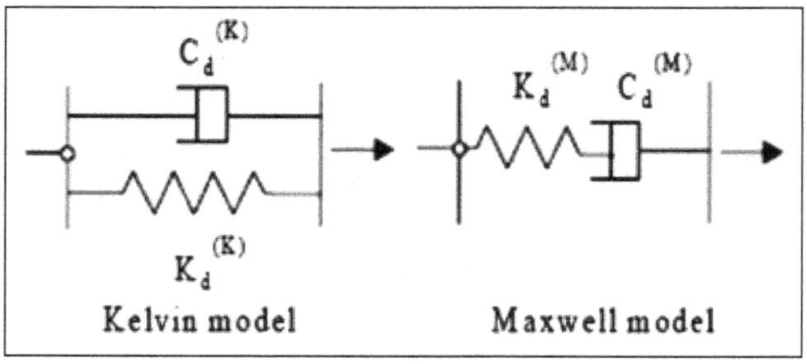

Maxwell model.

The model can also be described by the following equation:

$$P(t)+\lambda\frac{dP(t)}{dt}=C_d\cdot\frac{du_d}{dt}$$

where $u_d(t) = u_o \cdot \sin(\omega t)$. P is the damper output force, λ is the relaxation time, C_d is the damping constant at zero frequency, and u is the displacement of the piston head with respect to the damper housing. The relaxation time for the damper is defined as:

$$\lambda=\frac{C_d}{K_1}$$

where K_1 is the storage stiffness of the damper at infinite frequency.

For identification of the damper behaviour, the classical Maxwell model of $P(t)+\lambda\frac{dP(t)}{dt}=C_d\cdot\frac{du_d}{dt}$ was generalised to the following form in which the derivatives are of fractional order:

$$P(t)+\lambda\cdot D^r\left[P(t)\right]=C_d\cdot D^q\left[u(t)\right]$$

where D^r [P(t)] and D^q [u(t)] are fractional derivatives of orders *r* and *q*, which are based on material properties. For complex viscoelastic behaviour, the fractional derivative model typically offers an approved ability to describe the damper behaviour over a wide frequency range. Other more advanced models of viscoelasticity have been examined for modelling the behaviour of fluid damper.

Linear Fluid Viscous Damper

The model described by $P(t)+\lambda\cdot D^r\left[P(t)\right]=C_d\cdot D^q\left[u(t)\right]$ can be simplified to obtain a more useful model of linear viscous damping. When $r = q = 1$ the model becomes the Maxwell model described by $P(t)+\lambda\frac{dP(t)}{dt}=C_d\cdot\frac{du_d}{dt}$. The device parameters, λ and C_d, were obtained from experimental tests performed in studies. If the frequency of vibration is below the cut-off frequency, the second term in $P(t)+\lambda\cdot D^r\left[P(t)\right]=C_d\cdot D^q$ $\left[u(t)\right]$ drops out and the model of the damper can be simplified as:

$$P(t)+C_d\cdot D\frac{du_d}{dt}$$

where C_d is independent of the frequency, but dependent on ambient temperature.

The energy dissipated by damper is:

$$W_D=\oint F_D\cdot du$$
$$\Rightarrow W_D=\oint D_d\cdot u^\bullet\cdot du=\int_0^{2\cdot\pi/\omega}C_d\cdot u^{\bullet 2}\cdot dt$$
$$\Rightarrow W_D=C_d\cdot u_0^2\cdot\omega^2\int_0^{2\cdot\pi}C_d\cdot\cos^2(\omega\cdot t)\cdot d(\omega t)$$
$$\Rightarrow W_D=\pi\cdot C_d\cdot u_0^2\cdot\omega$$

Recognising that the damping ratio contributed by the damper can be expressed as $\xi_d = C_d/C_{cr}$ is obtained and natural excitation frequency is:

$$\omega_0=2\cdot\pi/T=\sqrt{K/M}$$
$$\Rightarrow W_D=\pi\cdot C_d\cdot u_0^2\cdot\omega=\pi\cdot\xi_d\cdot C_{er}\cdot u_0^2\cdot\omega$$
$$\Rightarrow W_D=2\cdot\pi\cdot\xi_d\cdot\sqrt{K\cdot M}\cdot u_0^2\cdot\omega$$
$$\Rightarrow W_D=2\cdot\pi\cdot\xi_d\cdot K\cdot u_0^2\cdot\frac{\omega}{\omega_0}$$
$$\Rightarrow W_D=4\cdot\pi\cdot\xi_d\cdot W_s\cdot\frac{\omega}{\omega_0}$$

where C_{cr}, K, M, ω_o and $W_s = K \cdot u_o^2/2$ are respectively the critical damping coefficient, stiffness, mass, natural frequency and elastic strain energy of the system.

The damping ratio attributed to the damper can then be expressed as,

$$\xi_d = \frac{W_D \cdot \omega_0}{4 \cdot \pi W_s \cdot \omega}$$

W_D and W_s are illustrated in figure. Under earthquake excitation, ω is essentially equal to ω_o and equation above is reduced to,

$$\xi_d = \frac{W_D}{4 \cdot \pi \cdot W_s}$$

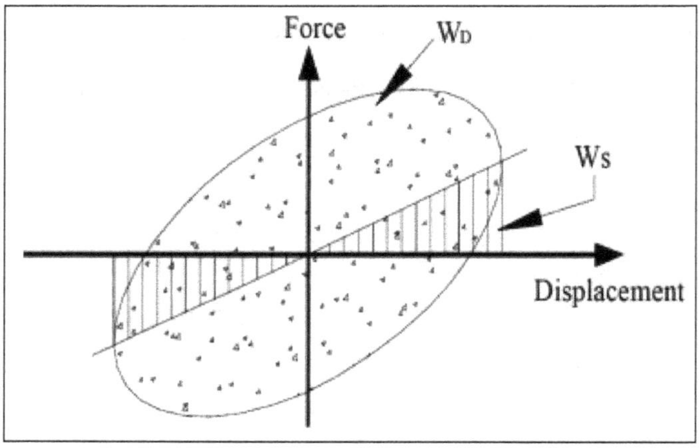

Definition of dissipation energy W_D and elastic strain energy of SDOF with FVD.

Modelling of System with Fluid Viscous Damper

Figure shows a structure with a multi degree of freedom connected to FVD. The motion equation of the structure subjected to a ground vibration becomes.

$$[M] \cdot U^{\bullet\bullet} + [(C)] \cdot U^{\bullet} + [(K)] \cdot U + F_d(t) = -[M] \cdot x_g^{\bullet\bullet}$$

Where

- M: Structure mass,
- K = Structure equivalent stiffness,
- C: Damping coefficient of the structure,
- F_d (t): FVD force vector,
- U, $U^{\bullet}$, $U^{\bullet\bullet}$: Displacement, velocity and acceleration vectors of the structure,
- $x_g^{\bullet\bullet}$: Ground acceleration (Earthquake).

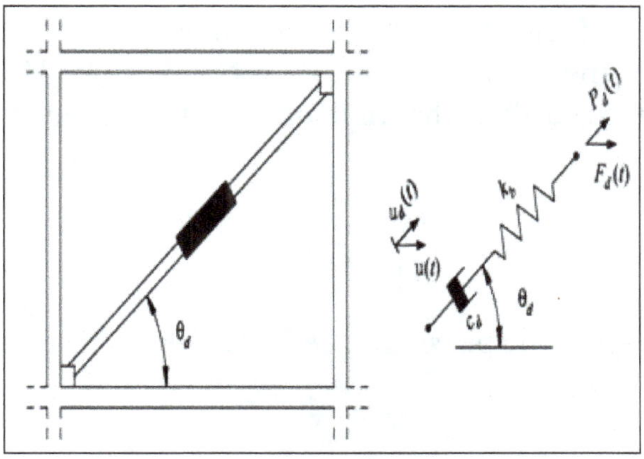

Multiple degree of freedom (MDOF) structure with FVD.

Considering a MDOF system as shown in figure the total effective damping ratio of the system ξ_{eff}, is defined as,

$$\xi_{eff} = \xi_0 + \xi_d$$

where ξ_o is the inherent damping ratio of the MDOF without dampers, and ξ_d is the damping ratio of the FVD. Extended from the concept of SDOF system, equation below is used by FEMA273 (Federal Emergency Management Agency) to represent ξ_d.

$$\xi_d = \frac{\sum W_j}{4 \cdot \pi \cdot W_K}$$

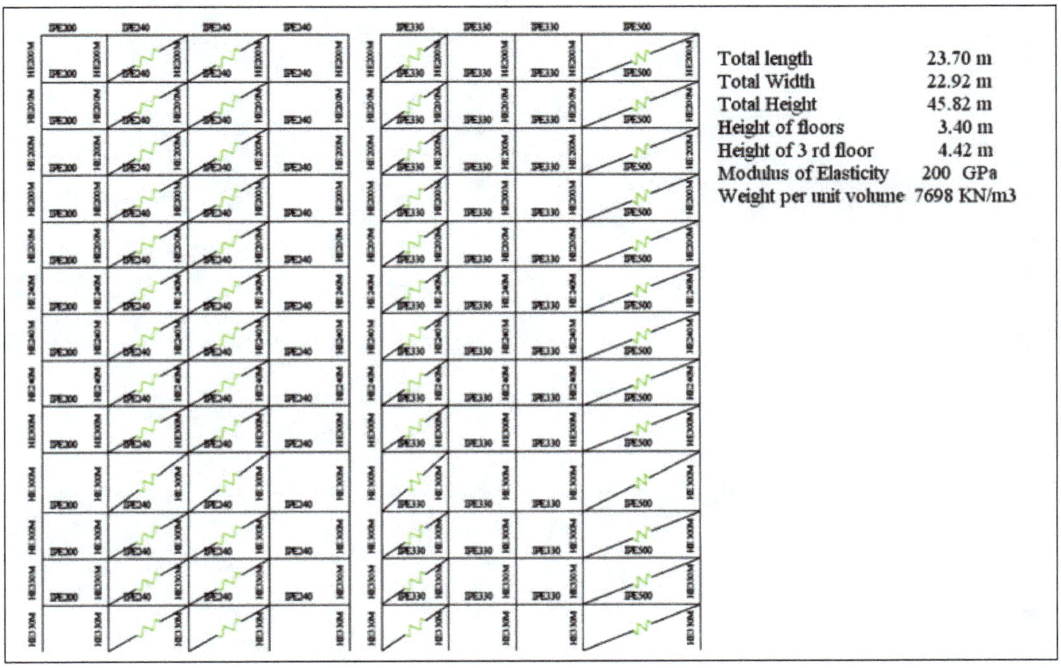

Modelling of twelve-storey building connected to FVD.

$\sum W_j$ is the sum of the energy dissipated by the *j*th damper of the system in one cycle; W_K is the elastic strain energy of the frame. W_K is equal to $\sum F_i \cdot U_i$ where Fi is the storey shear and U_i is the storey drift of the *i*th floor. Now the energy dissipated by the FVD can be expressed by,

$$\sum_j W_j = \sum_j \pi \cdot C_j \cdot u_j^2 \cdot \omega_0 = \frac{2 \cdot \pi^2}{T} \cdot \sum_j C_j \cdot u_j^2$$

where u_j is the relative axial displacement between two ends of the damper.

References

- Products-and-solutions.seismic-protection, fpc-italia-en.nsf, freyssinet: freyssinet.com, Retrieved 16 May, 2020
- Riahi, H. T. and Estekanchi, H. E, (2010), "Seismic Assessment of Steel Frames with Endurance Time Method", Journal of Constructional Steel Research, v66, n6, p780-792, doi:10.1016/j.jcsr.2009.12.001
- Reitherman, Robert (2012). Earthquakes and Engineers: An International History. Reston, VA: ASCE Press. pp. 486–487. ISBN 9780784410714. Archived from the original on 2012-07-26
- Seismic-risk-of-structures-and-the-economic-issues-of-earthquakes, earthquake-engineering: intechopen.com, Retrieved 6 July, 2020
- Brun, M.; S. Guenneau; and A.B. Movchan (2009-02-09). "Achieving control of in-plane elastic waves". Appl. Phys. Lett. 94 (61903): 061903. arXiv:0812.0912. Bibcode:2009ApPhL..94f1903B. doi:10.1063/1.3068491

Permissions

All chapters in this book are published with permission under the Creative Commons Attribution Share Alike License or equivalent. Every chapter published in this book has been scrutinized by our experts. Their significance has been extensively debated. The topics covered herein carry significant information for a comprehensive understanding. They may even be implemented as practical applications or may be referred to as a beginning point for further studies.

We would like to thank the editorial team for lending their expertise to make the book truly unique. They have played a crucial role in the development of this book. Without their invaluable contributions this book wouldn't have been possible. They have made vital efforts to compile up to date information on the varied aspects of this subject to make this book a valuable addition to the collection of many professionals and students.

This book was conceptualized with the vision of imparting up-to-date and integrated information in this field. To ensure the same, a matchless editorial board was set up. Every individual on the board went through rigorous rounds of assessment to prove their worth. After which they invested a large part of their time researching and compiling the most relevant data for our readers.

The editorial board has been involved in producing this book since its inception. They have spent rigorous hours researching and exploring the diverse topics which have resulted in the successful publishing of this book. They have passed on their knowledge of decades through this book. To expedite this challenging task, the publisher supported the team at every step. A small team of assistant editors was also appointed to further simplify the editing procedure and attain best results for the readers.

Apart from the editorial board, the designing team has also invested a significant amount of their time in understanding the subject and creating the most relevant covers. They scrutinized every image to scout for the most suitable representation of the subject and create an appropriate cover for the book.

The publishing team has been an ardent support to the editorial, designing and production team. Their endless efforts to recruit the best for this project, has resulted in the accomplishment of this book. They are a veteran in the field of academics and their pool of knowledge is as vast as their experience in printing. Their expertise and guidance has proved useful at every step. Their uncompromising quality standards have made this book an exceptional effort. Their encouragement from time to time has been an inspiration for everyone.

The publisher and the editorial board hope that this book will prove to be a valuable piece of knowledge for students, practitioners and scholars across the globe.

Index

www.ingramcontent.com/pod-product-compliance
Lightning Source LLC
LaVergne TN
LVHW080610210926
848485LV00003B/133